T0256203

MASSEY FERGUSON

SHOP MANUAL

More information available at haynes.com
Phone: 805-498-6703

Haynes Group Limited
Haynes North America, Inc.

ISBN-10: 0-87288-520-8
ISBN-13: 978-0-87288-520-2

Information and Instructions

This individual Shop Manual is one unit of a series on agricultural wheel-type tractors. Contained in it are the necessary specifications and the brief but terse procedural data needed by a mechanic when repairing a tractor on which he has had no previous actual experience.

The material is arranged in a systematic order beginning with an index which is followed immediately by a Table of Condensed Service Specifications. These specifications include dimensions, clearances, capacities and tune-up information. Next in order of arrangement is the procedures section.

In the procedures section, the order of presentation starts with the front axle system and steering and proceeds toward the rear axle. The last portion of the procedures section is devoted to the power take-off and power lift systems. Interspersed where needed in this section are additional tabular specifications pertaining to wear limits, torquing, etc.

How to use the index

Suppose you want to know the procedure for R&R (remove and reinstall) of the engine camshaft. Your first step is to look in the index under the main heading of "Engine" until you find the entry "Camshaft." Now read to the right. Under the column covering the tractor you are repairing, you will find a number which indicates the beginning paragraph pertaining to the camshaft. To locate this paragraph in the manual, turn the pages until the running index appearing on the top outside corner of each page contains the number you are seeking. In this paragraph, you will find the information concerning the removal of the camshaft.

Common spark plug conditions

NORMAL

Symptoms: Brown to grayish-tan color and slight electrode wear. Correct heat range for engine and operating conditions.
Recommendation: When new spark plugs are installed, replace with plugs of the same heat range.

WORN

Symptoms: Rounded electrodes with a small amount of deposits on the firing end. Normal color. Causes hard starting in damp or cold weather and poor fuel economy.
Recommendation: Plugs have been left in the engine too long. Replace with new plugs of the same heat range. Follow the recommended maintenance schedule.

TOO HOT

Symptoms: Blistered, white insulator, eroded electrode and absence of deposits. Results in shortened plug life.
Recommendation: Check for the correct plug heat range, over-advanced ignition timing, lean fuel mixture, intake manifold vacuum leaks, sticking valves and insufficient engine cooling.

CARBON DEPOSITS

Symptoms: Dry sooty deposits indicate a rich mixture or weak ignition. Causes misfiring, hard starting and hesitation.
Recommendation: Make sure the plug has the correct heat range. Check for a clogged air filter or problem in the fuel system or engine management system. Also check for ignition system problems.

PREIGNITION

Symptoms: Melted electrodes. Insulators are white, but may be dirty due to misfiring or flying debris in the combustion chamber. Can lead to engine damage.
Recommendation: Check for the correct plug heat range, over-advanced ignition timing, lean fuel mixture, insufficient engine cooling and lack of lubrication.

ASH DEPOSITS

Symptoms: Light brown deposits encrusted on the side or center electrodes or both. Derived from oil and/or fuel additives. Excessive amounts may mask the spark, causing misfiring and hesitation during acceleration.
Recommendation: If excessive deposits accumulate over a short time or low mileage, install new valve guide seals to prevent seepage of oil into the combustion chambers. Also try changing gasoline brands.

HIGH SPEED GLAZING

Symptoms: Insulator has yellowish, glazed appearance. Indicates that combustion chamber temperatures have risen suddenly during hard acceleration. Normal deposits melt to form a conductive coating. Causes misfiring at high speeds.
Recommendation: Install new plugs. Consider using a colder plug if driving habits warrant.

OIL DEPOSITS

Symptoms: Oily coating caused by poor oil control. Oil is leaking past worn valve guides or piston rings into the combustion chamber. Causes hard starting, misfiring and hesitation.
Recommendation: Correct the mechanical condition with necessary repairs and install new plugs.

DETONATION

Symptoms: Insulators may be cracked or chipped. Improper gap setting techniques can also result in a fractured insulator tip. Can lead to piston damage.
Recommendation: Make sure the fuel anti-knock values meet engine requirements. Use care when setting the gaps on new plugs. Avoid lugging the engine.

GAP BRIDGING

Symptoms: Combustion deposits lodge between the electrodes. Heavy deposits accumulate and bridge the electrode gap. The plug ceases to fire, resulting in a dead cylinder.
Recommendation: Locate the faulty plug and remove the deposits from between the electrodes.

MECHANICAL DAMAGE

Symptoms: May be caused by a foreign object in the combustion chamber or the piston striking an incorrect reach (too long) plug. Causes a dead cylinder and could result in piston damage.
Recommendation: Repair the mechanical damage. Remove the foreign object from the engine and/or install the correct reach plug.

SHOP MANUAL
MASSEY-FERGUSON

MODELS

362   365   375   383   390   390T   398

The tractor serial number is located in two locations, on the serial number plate (Fig. 1) located on the right side of the tractor and stamped on the rear axle casting (Fig. 2). The tractor serial number is coded to identify the machine type, year of manufacture, week of manufacture (during specific year) and specific unit during week.

The illustration shown in Fig. 3 identifies a two-wheel-drive 390 tractor which was the 121st tractor manufactured during the 32nd week of 1988.

Fig. 1—The tractor serial number is located on plate attached to tractor right side as shown as well as stamped on axle housing as shown in Fig. 2.

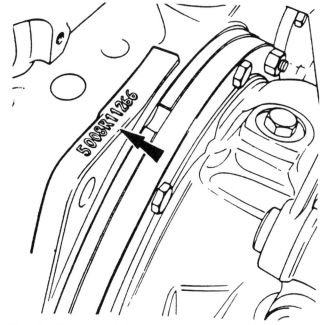

Fig. 2—Tractor serial number is stamped into rear axle housing as shown as well as on plate shown in Fig. 1. Numbers match on original assembly.

Year of manufacture   Tractor Built In Week Of

5008N32121

Machine Type   Week Of Manufacture

Year of manufacture –
A = Jan. 1992 - Dec. 1992
B = Jan. 1993 - Dec. 1993
N = Feb. 1988 - Jan. 1989
P = Feb. 1989 - Jan. 1990

R = Feb. 1990 - Dec. 1990
S = Jan. 1991 - Dec. 1991
U = Aug. 1986 - Jan. 1987
V = Feb. 1987 - Jan. 1988

Fig. 3—The tractor serial number identifies machine type, year of manufacture, week of manufacture during specific year, and specific unit manufactured during that week. The example shown identifies a two-wheel-drive 390 tractor which was the 121st tractor manufactured during the 32nd week of 1988.

Machine type –
5006 M-F 375 2WD	5270 M-F 365 2WD
5007 M-F 375 4WD	5271 M-F 365 4WD
5008 M-F 390 2WD	5742 M-F 383 4WD
5009 M-F 390 4WD	5723 M-F 390T 2WD
5010 M-F 398 2WD	5724 M-F 390T 4WD
5011 M-F 398 4WD	5726 M-F 362 2WD
5266 M-F 383 2WD	5727 M-F 362 4WD

INDEX (By Starting Paragraph)

DUAL DIMENSIONS

This service manual provides specifications in both U.S. Customary and Metric (SI) systems of measurement. The first specification is given in the measuring system perceived by us to be the preferred system when servicing a particular component, while the second specification (given in parenthesis) is the converted measurement. For instance, a specification of 0.011 inch (0.28 mm) would indicate that we feel the preferred measurement in this instance is the U.S. Customary system of measurement and the Metric equivalent of 0.011 inch is 0.28 mm.

CONDENSED SERVICE DATA

			Models		
	362	365	375	383-390	390T-398
GENERAL					
Engine Make			Perkins		
Model		A4.236		A4.248S	AT4.236
Build Code	LD31234	LD31190	LD31140	LF31141	LJ31142
Number of Cylinders			4		
Bore		98.4 mm (3.875 in.)		101 mm (3.975 in.)	98.4 mm (3.875 in.)
Stroke			127 mm (5.0 in.)		
Displacement		3.86 L (236 cid)		4.07 L (248 cid)	3.86 L (236 cid)
Compression Ratio			16:1		15.5:1
Firing Order			1-3-4-2		

CONDENSED SERVICE DATA (CONT.)

	362	365	Models 375	383-390	390T-398
GENERAL (Cont.)					
Valve Clearance (Cold) –					
Inlet			0.3 mm (0.012 in.)		
Exhaust			0.3 mm (0.012 in.)		
Valve Face Angle					
Inlet			45°		30°
Exhaust			45°		
Valve Seat Angle					
Inlet			45°		30°
Exhaust			45°		
Injection timing – BTDC					
Static			23°	24°	16°
Fuel Pump					
Make			CAV		
Model			DPA		
Engine Low Idle, rpm			725-775		
Engine High Idle, rpm	2310	2420*	2420*	2380*	2420*
Engine Rated Speed, rpm			2200		
Battery Terminal Grounded			Negative		

* High idle no load speed should be 2310 rpm for M-F 365 after engine serial number U261448S, M-F 375 after S. N. U209015P, M-F 390 after S. N. U226834S and M-F 390T/398 after S.N. U219831S.

SIZES

Crankshaft Main Journal Diameter	See Paragraph 57
Crankshaft Crankpin Diameter	See Paragraph 57
Camshaft Journal Diameter	See Paragraph 52
Piston Pin Diameter	See Paragraph 54
Valve Stem Diameter	See Paragraph 38

CLEARANCES

Main Bearing Diametral Clearance	See Paragraph 57
Rod Bearing Diametral Clearance	See Paragraph 56
Camshaft Bearing Diametral Clearance	See Paragraph 52
Crankshaft End Play	See Paragraph 57
Piston Skirt to Cylinder Clearance	See Paragraph 54

CAPACITIES

	362	375	390T-398
Cooling System	14.4 L (15.2 qts.)	15.1 L (16 qts.)	15.5 L (16.4 qts.)
Crankcase With Filter	6.8 L (7.2 qts.)	7.5 L (7.9 qts.)	7.7 L (8.1 qts.)
Transmission – Without Spacer		43.4 L (11.5 gal.)	

CONDENSED SERVICE DATA (CONT.)

	Models				
	362	365	375	383-390	390T-398

CAPACITIES (Cont.)
Transmission (Cont.)
  With Spacer or
    Transfer Housing . . . . . . ──────────────── 47.4 L ────────────────
                                   (12.5 gal.)

Final Drive Rear
  Planetary Hub – Each Side . ──────────────── 0.9 L ────────────────
                                   (5 pints)

Hydrostatic Steering . . . . . . ──────────────── 1.2 L ────────────────
                                   (2.2 pints)

Front Drive Axle Hubs
  (Each Side) . . . . ──────────── 1.2 L ────────────   1.3 L†
                             (2.3 pints)   (2.6 pints)

Front Drive
  Axle Housing . . . . 4.0 L ────────── 5.0 L ────────── 5.8 L‡
                (1.1 gal.)   (1.3 gal.)   (1.5 gal.)

† Capacity for 390T model is 1.2 L (2.3 pints).
‡ Capacity for 390T model is 5.0 L (1.3 gal.).

FRONT AXLE SYSTEM
(TWO-WHEEL-DRIVE)

FRONT AXLE ASSEMBLY AND STEERING LINKAGE

Two-Wheel-Drive Models

1. WHEELS AND BEARINGS. To remove front wheel hub and bearings, first raise and support the front axle extension, then unbolt and remove the tire and wheel assembly. Remove cap (2 or 3—Fig. 4), cotter pin (4), castellated nut (5), washer and outer bearing cone (7). Slide the hub assembly from spindle axle shaft. Remove dust shield (12), seal (9) and inner bearing cone (11). Drive bearing cups (8 and 10) from hub if renewal is required. Pack wheel bearings liberally with a multi-purpose lithium based grease. Reassemble by reversing disassembly procedure. Tighten castellated nut (5) to a torque of 80 N·m (60 ft.-lbs.), then back nut off to the nearest hole and install cotter pin (4). Be sure to install cap (2 or 3) securely.

2. TRACK ROD AND TOE-IN. All models are equipped with hydrostatic steering. On 362 models, a single track rod connects the left and right steering arms which are attached to the steering spindles. On 365-398 models, the two track rods are attached to each end of the hydrostatic steering rod, which is located between the steering arms. The track rod of all models assures that both left and right wheels turn in unison and the distance between ends of track rod establishes front wheel toe-in.

Ends of track rod are automotive type and should be renewed if wear is excessive. The procedure for removing and installing ends is self-evident. Recommended toe-in is 0-5 mm (0-3/16 inch) for all models. Toe-in should be measured between the wheel rims on center line of axle, parallel to ground, at front and rear of wheels. Rotate wheels and remeasure to be sure that wheels are not bent giving incorrect reading.

Front wheel tread width is adjustable to seven different widths by relocating axle extensions and changing length of track rod using the preexisting attachment holes.

On 362 models, axle extension and hydrostatic steering cylinder bracket retaining screws (1—Fig. 5) should be tightened to 180-230 N·m (135-170 ft.-lbs.) torque. Tighten track setting screw (2) to 45-55 N·m (33-40 ft.-lbs.) torque before tightening locknut. Note that spacers are used on the outer screws between steering cylinder brackets and axle extensions when axle width is at the four widest settings. To make small toe-in adjustments, remove track adjusting screw (2) from left end of track rod, loosen rod end clamp bolt on right end, then turn center section of track rod until toe-in is correct. Reinstall track adjusting screw (2) and tighten rod end clamp bolt to 45-55 N·m (33-41 ft.-lbs.) torque.

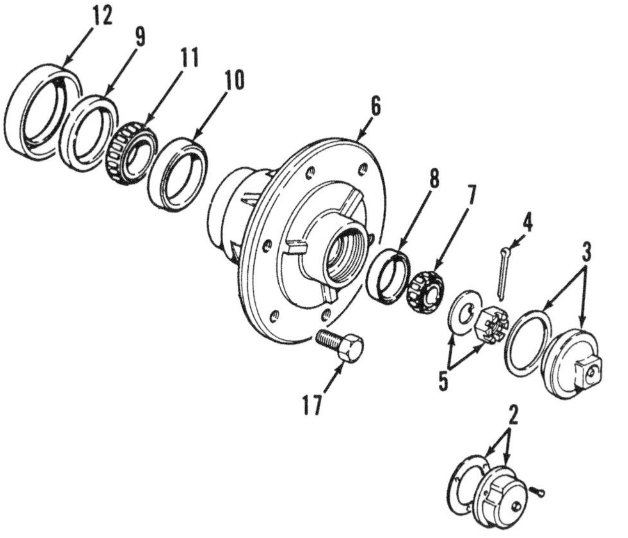

Fig. 4—Exploded view of wheel hub typical of two-wheel-drive models.

2. Hub cap	
3. Hub cap	8. Outer bearing cup
4. Cotter pin	9. Seal
5. Castellated nut	10. Inner bearing cup
& tang washer	11. Inner bearing cone
6. Hub	12. Dust shield
7. Outer bearing cone	17. Wheel retaining screws

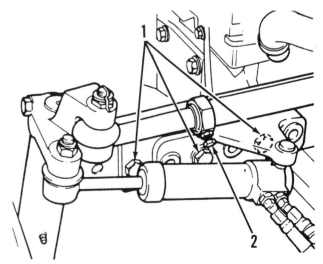

Fig. 5—View of axle left end typical of 362 model. Track adjusting screw and locknut (2) and cylinder bracket and axle extension screws (1) must be properly installed and tightened.

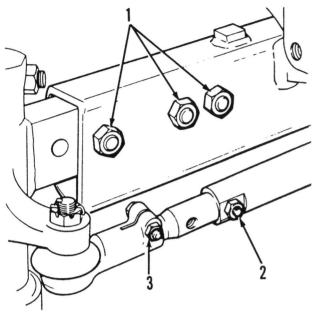

Fig. 6—View of axle left end typical of type used on 365-398 models. Right side is equipped with a similar track rod, adjusting bolt (2) and rod end. Rod end may be locked with clamp bolt (3) as shown or locknut.

On 365-398 models, axle extension retaining screws (1—Fig. 6) should be tightened to 340-450 N·m (250-330 ft.-lbs.) torque. Tighten track setting bolt (2) to 120-160 N·m (90-120 ft.-lbs.). To make small toe-in adjustments, remove track adjusting bolt (2), loosen rod end clamp bolt (3) or jam nut, then turn track rod until toe-in is correct. Reinstall track adjusting bolt (2) and tighten to 120-160 N·m (90-120 ft.-lbs.) torque. Tighten rod end clamp bolt (3) to 45 N·m (33 ft.-lbs.) torque or jam nut to 160-200 N·m (120-130 ft.-lbs.) torque. Equal toe-in adjustments should be made to both sides to center steering.

3. SPINDLES, AXLE EXTENSIONS AND BUSHINGS. To remove spindle (12—Fig. 7 or Fig. 8), first remove the wheel and hub. Disconnect rod end from steering arm (1), remove clamp screw (13) from steering arm, then remove steering arm. Remove key (14) and seal (20) from top of spindle, then lower spindle out of axle extension (19). Remove thrust bearing (11) from spindle. Clean and inspect parts for wear or other damage and renew as necessary.

Each axle extension (19) is equipped with two renewable spindle bushings (18). New bushings must be reamed after they are pressed into position. Clean all metal particles from bore and be sure that hole for grease fitting is clean and open before assembling.

When reassembling, install thrust bearing (11) on spindle so that numbered side of bearing is facing upward and insert spindle through axle extension. Install seal (20) and key (14) then locate steering arm on top of spindle. Tighten steering arm retaining clamp screw to a torque of 125-165 N·m (94-122 ft.-lbs.) for 362 models; 280-370 N·m (207-273 ft.-lbs.) torque for 365-398 models. Refer to paragraph 2 for track and toe-in adjustment and other recommended torques. Balance of reassembly is the reverse of disassembly.

4. AXLE CENTER MEMBER, PIVOT PIN AND BUSHINGS. To remove front axle assembly, first remove any front mounted equipment, guards, weights and weight frame. Raise front of tractor in such a way that it will not interfere with the removal of the axle, such as with a support located under engine sump. Removal of wheels, spindles and axle extensions will reduce weight and may make handling the center member easier; however, the complete axle assembly can be removed as a unit.

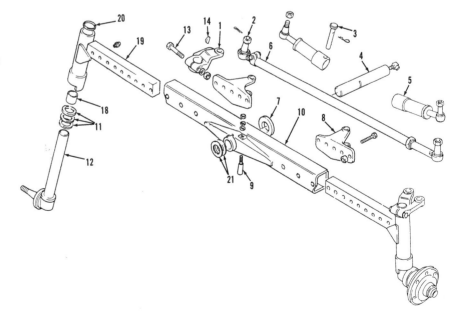

Fig. 7—Exploded view of front axle typical of two-wheel-drive 362 model.

1. Steering arm
2. Rod end
3. Cylinder pivot pin
4. Axle pivot pin
5. Steering cylinder
6. Track rod
7. Thrust washer
8. Steering cylinder bracket
9. Tapered retaining pin
10. Axle center member
11. Thrust washers
12. Spindle
18. Bushing
19. Axle extension
20. Seal
21. Shims

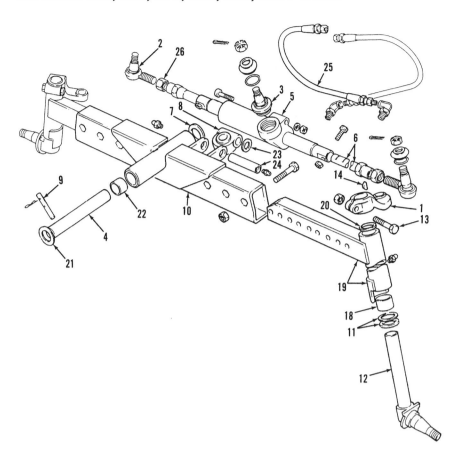

Fig. 8—Exploded view of front axle typical of type used on two-wheel-drive 365-398 models.

1. Steering arm
2. Rod end
3. Cylinder ball joint
4. Axle pivot pin
5. Steering cylinder
6. Track rod
7. Thrust washer
8. Pivot block
9. Retaining pin
10. Axle center member
11. Thrust washers
12. Spindle
14. Key
18. Bushing
19. Axle extension
20. Seal
21. Shims
22. Bushing
23. Shims
24. Pivot pin
25. Hydraulic hoses
26. Jam nut

Disconnect hydrostatic steering hoses from the steering cylinder or cylinders and cover openings to prevent the entry of dirt. Support the axle with a suitable jack to prevent tipping while permitting the axle to be lowered and moved safely. Remove retaining pin (9—Fig. 7 or Fig. 8), then use a suitable puller to withdraw axle pivot pin (4). Carefully lower the axle assembly and roll axle from under tractor.

Check axle pivot bushings and renew if necessary. Bushings are located in support housing of 362 models and should be installed flush to 0.5 mm (0.020 inch) below flush (A—Fig. 9) with housing bore. Split (C—Fig. 9) in bushing should be down and hole (C) for grease passage should be up as shown. Axle pivot bushings are located in axle of 365-398 models. On all models, it may be necessary to ream bushings after installation. Reverse removal procedure when assembling. Axle end play should be 0.05-0.25 mm (0.002-0.010 inch) on pivot pin. Push the axle toward rear on pivot pin, then measure axle end play with a feeler gauge as shown in Fig. 10 or Fig. 11. Shims (21—Fig. 7 or Fig. 8) are available in various thicknesses for adjusting end play. On 362 models, make sure that tapered pin (9—Fig. 7) is correctly and firmly seated, then tighten retaining nut to 80-140 N·m (70-110 ft.-lbs.) torque. On all models, refer to paragraphs 2 and 3 for additional torque values and assembly notes.

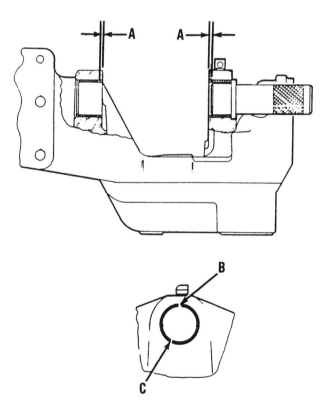

Fig. 9—Axle pivot bushings should be installed as shown for 362 model. Hole (B) should be aligned with grease passage and slot (C) should be in position indicated.

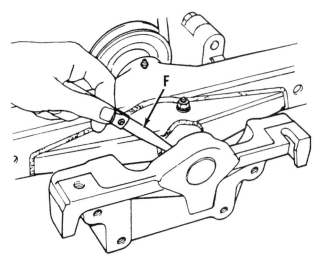

Fig. 10—End play of front axle should be measured with a feeler gauge as shown for 362 model with two-wheel drive.

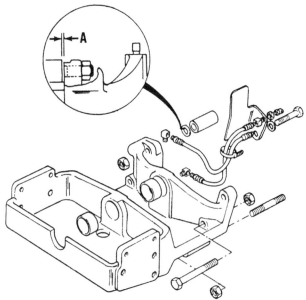

Fig. 12—Measure gap (A) between the top spacer tube on right side and front support of 362 model.

assembly, then refer to paragraph 4 and remove the axle assembly. Attach a hoist or other supporting device to the front support, then unbolt and separate the front support from the front of the engine. Be careful not to lose shims which may be located between front support casting and front of engine.

To remove front axle support from 365-398 models, first remove any front mounted equipment, guards,

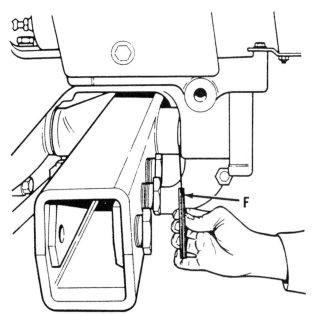

Fig. 11—Measure end play of front axle with a feeler gauge as shown for models 365-398 with two-wheel drive.

5. FRONT SUPPORT. To remove the front support, the axle must be removed, the radiator must be removed and the front support must be unbolted from the front of engine. The front axle, the front support and the remainder of the tractor must each be supported separately while removing, while separated and while assembling. Be sure that sufficient equipment is available before beginning.

To remove front axle support from 362 models, first remove any front mounted equipment, guards, weights and weight frame. Remove grille, hood, hood side panels, air cleaner and battery. Drain cooling system, disconnect radiator hoses and disconnect wires to headlights. Remove radiator and oil cooler

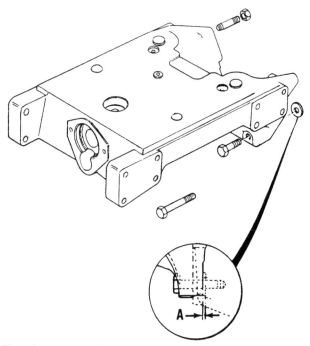

Fig. 13—Use a feeler gauge to measure gap (A) between the lower ears of casting and four-cylinder engine of 365-398 models.

weights and weight frame. Remove fuel tank and radiator and radiator support frame, then refer to paragraph 4 and remove the axle assembly. Attach a hoist or other supporting device to the front support, then unbolt and separate the front support from the front of the engine. Be careful not to lose shims which may be located between front support casting and front of engine.

Reattach front support to engine of all models by reversing the removal procedure, but omitting any shims originally installed between support and engine. On 362 models, tighten the retaining screws to 240-320 N•m (177-236 ft.-lbs.) torque. Measure any

gap (A—Fig. 12) between the top spacer tube on right side and front support with a feeler gauge. Loosen retaining screws and install shims equal to the measured gap plus 0.13 mm (0.005 inch).

Tighten screws retaining front support of 365-398 models to 230-255 N•m (170-190 ft.-lbs.) torque. Use a feeler gauge to measure any gap (A—Fig. 13) between the lower ears of front support casting and engine of 365-398 models. Loosen retaining screws and install shims equal to the measured gap plus 0.13 mm (0.005 inch).

On all models, complete assembly by reversing the removal procedure.

FRONT-WHEEL DRIVE

7. A mechanical front-wheel drive is available on these models. There are some differences between the Front-Wheel Drive Systems used on these models that will be referred to in the servicing instructions which follow.

The front drive is engaged by an electric solenoid/hydraulic valve which directs oil pressure to move a dog clutch, or by a dog clutch which is moved manually by a shift fork by way of mechanical linkage. The transfer gearbox is attached to the left side of the range gearbox of models with twelve-speed shuttle transmissions or located between the rear of the transmission housing and the front of the rear axle housing of other models. On all models, a drive shaft with two "U" joints connects the transfer gearbox to front axle.

TRACK ROD AND TOE-IN

All Four-Wheel-Drive Models

8. All models are equipped with hydrostatic steering. A single track rod connects the left and right steering arms which are attached to the steering spindles. The track rod assures that both left and right wheels turn in unison, and the distance between ends of track rod establishes front wheel toe-in.

Ends of track rod are automotive type and should be renewed if wear is excessive. The procedure for removing and installing ends is self-evident. Recommended toe-in is 0 for all models. Toe-in should be measured between the wheel rims on center line of axle, parallel to ground, at front and rear of wheels. Rotate wheels and remeasure to be sure that wheels are not bent giving incorrect reading.

Front wheel tread width is adjustable to different widths by relocating the wheel on the center disc or by reversing the wheels. If wheels are reversed, they

must be moved to opposite sides of tractor to maintain correct tire tread direction.

To adjust toe-in, loosen the locknuts at each end of the track (tie) rod, then turn the tie rod tube to set the toe-in. Tighten locknuts at each end when adjustment is correct. Nut retaining ball-joint of track rod end in the steering arm should be tightened to 78-86 N•m (58-63 ft.-lbs.) torque for 362-390T models, 108-118 N•m (80-87 ft.-lbs.) torque for 398 models.

DRIVE SHAFT

All Models So Equipped

9. REMOVE AND REINSTALL. To remove drive shaft, first loosen clamps (1 and 2—Fig. 14), then slide guard (3) into the center tube. Remove spring clip (6) from guard at front. Remove bolts (4 and 5) from ends, then remove the shield and drive shaft assembly. Unscrew seal retaining ring from the sliding coupling, remove sliding coupling from rear end of drive shaft, then remove drive shaft from guard tube.

When reassembling, grease splines of sliding coupling and both universal joints. Insert the long section of drive shaft into guard, then assemble sliding coupling at rear, making sure that arrows on drive shaft and sliding coupling are aligned. Coat threads of bolts (4 and 5) with "Loctite 270," then attach drive shaft flanges with bolts (4 and 5) tightened to 55-75 N•m (40-55 ft.-lbs.) torque. Install clip (6) at front (axle end), making sure that drain hole in guard is toward bottom. Distance from step in axle and front of guard should be approximately 90 mm (3.5 inch).

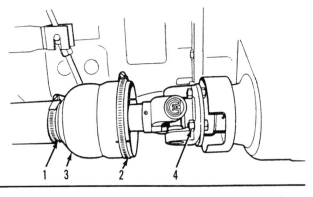

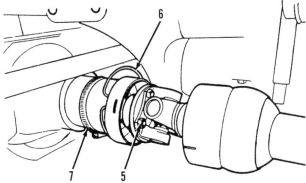

Fig. 14—View showing both ends of drive shaft and shield. Front (axle) end of shield is retained by clip (6).

1. Clamp
2. Clamp
3. Guard
4. Retaining bolts
5. Retaining bolts
6. Clip

FRONT DRIVE AXLE

All Models So Equipped

10. R&R ASSEMBLY. First remove the drive shaft and shield as outlined in paragraph 9. Raise front of tractor in such a way that it will not interfere with the removal of the axle. Remove front wheels and weights, then support the axle with a suitable jack or special safety stand to prevent tipping while permitting the axle to be lowered and moved safely. Disconnect hoses from the steering cylinder and cover openings to prevent the entry of dirt. Remove retaining pin or screw from the front pivot, then use a suitable puller to withdraw axle pivot pin. Lower axle until free, then carefully roll axle away.

Reinstall front drive axle by reversing the removal procedure. A spacer is located at the rear of the axle pivot and sufficient thickness of shims should be located in front of axle to reduce end play to 0.05-0.25 mm (0.002-0.010 inch). Grease pivot after correct thickness of shims and pivot pin retaining screw or pin have been installed.

PLANETARY ASSEMBLY

All Models So Equipped

11. R&R AND OVERHAUL. Refer to Fig. 15. Either planetary assembly can be serviced without removing the steering knuckle housing (26—Fig. 16). However, if seal (24) or bearings (20 and 23) are to be serviced, hub (22) and knuckle housing (26) must first be removed as outlined in paragraph 12. Support front axle housing and remove front wheel. Rotate hub (22) until drain plug (1), is down, then remove plug and drain oil from hub assembly. Remove screws (2) and lift off planetary carrier (3). Remove snap ring (12), then bump shafts (10) from housing bores. Keep each planet assembly, consisting of shaft (10), gear (7), bearings (8) and thrust washers (6), separate so

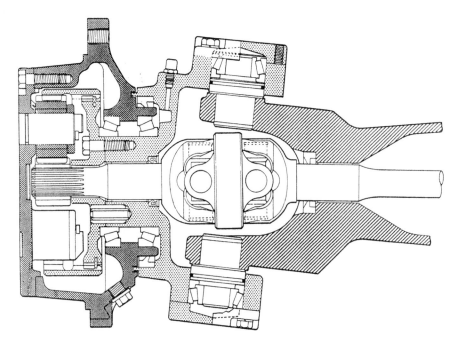

Fig. 15—Cross section of typical front drive axle.

that they can be reinstalled as a set. The sun gear (13) can be removed after removing snap ring (9).

Clean and inspect all parts for excessive wear or other damage and renew as necessary.

Install snap rings (11), bearings (8), planetary gears (7) and thrust washers (6), then secure with snap ring (12). Opening in snap ring (12) must be positioned in center of one of the three blocks, not in an open area. Make sure thrust plate (5) is in place. Clean mating surfaces of wheel hub (22) and planet carrier (3), then coat face with "Loctite 515 Instant Gasket" and install carrier assembly. Coat screws (2)

lightly with oil before installing, then tighten to 96-118 N·m (71-87 ft.-lbs.) torque.

Turn wheel hub (22) until hole for plug (1) is horizontal, fill hub and planetary to the level of opening for plug (1) with an approved oil, then install and tighten plug. Recommended oil for use in front drive axle is Massey-Ferguson Super 500 multi-use 10W-30 oil, Massey-Ferguson Permatran or equivalent. Install front wheel and tighten disc to hub nuts to 270 N·m (200 ft.-lbs.) torque. Disc to wheel rim nuts should be tightened to 190 N·m (140 ft.-lbs.) torque.

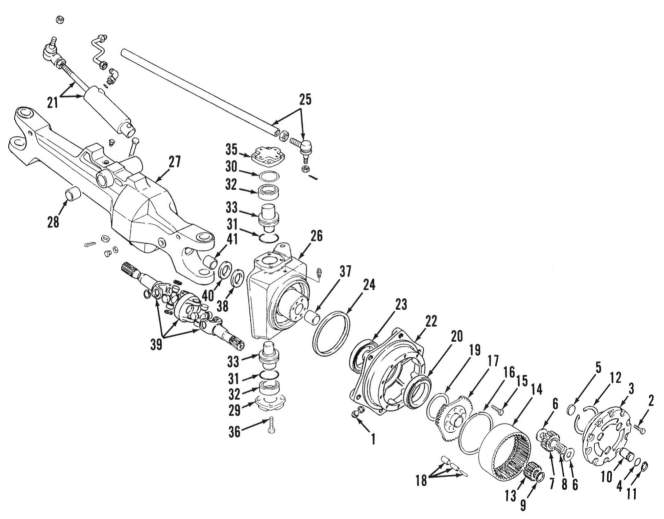

Fig. 16—Exploded view of front-wheel-drive steering knuckle and planetary assembly.

1. Drain plug	12. Snap ring	22. Wheel hub	31. "O" ring
2. Screws	13. Sun gear	23. Bearing cup and cone	32. Lower bearing
3. Planet carrier	14. Ring gear	24. Seal	33. Pivot pin
4. "O" ring	15. Screws (Metric)	25. Track rod and rod end	35. Upper retainer
5. Thrust pin	16. Snap ring	26. Steering knuckle	36. Screw
6. Thrust washer	17. Ring gear hub	27. Axle center housing	37. Bushing
7. Planet gear	18. Pins	28. Bushing	38. Seal
8. Rollers (50/gear)	19. Shims (0.10, 0.15, 0.30,	29. Lower retainer	39. Axle shaft assy.
9. Snap ring	0.50, 0.70, 1.00 mm)	30. Shims (0.10, 0.15, 0.20,	40. Seal
10. Planet gear shaft	20. Bearing cup and cone	0.50, 1.00 mm)	41. Bushing
11. Snap ring	21. Steering cylinder		

STEERING KNUCKLE HOUSING AND WHEEL HUB

All Models So Equipped

12. R&R AND OVERHAUL. To remove either steering knuckle housing, first remove planetary as outlined in paragraph 11. Remove snap ring (9—Fig. 16) and sun gear (13). Disconnect steering cylinder (21) from right side steering knuckle arm, and disconnect track rod (25) from steering knuckle arms on both sides. Unbolt and remove upper and lower retainers (29 and 35) and shims (30). Measure and note thickness and number of shims (30) under each retainer for aid in reassembly. Carefully remove the bearing cups for upper and lower bearings (32). Use a suitable puller such as MF 451A or equivalent to pull upper and lower pivot pins (33). Carefully remove steering knuckle housing (26) and wheel hub assembly (22) from axle center housing (27), leaving axle shafts (39) with center housing.

Axle shaft and double "U" joint assembly (39) may be withdrawn from center housing for inspection or repair. If renewal is required, oil seal (38) and bushing (37) can be removed from knuckle housing (26) and oil seal (40) and bushing (41) can be removed from axle housing. Bushings (37 and 41) should be pressed into position with external groove toward top and internal arrow-shaped grooves pointing toward in-

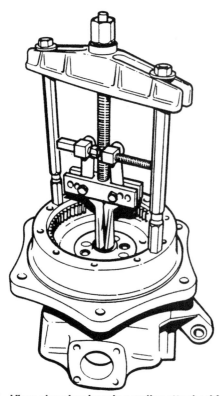

Fig. 17—View showing bearing puller attached for pulling the ring gear from steering knuckle. Refer to text.

side of oil-filled housing (away from seal). Be careful not to damage seals (38 and 40) when installing axle and knuckle housing.

To disassemble the wheel hub and steering knuckle, remove the six cap screws (15). Attach an internal bearing puller as shown in Fig. 17 and pull ring gear hub (17—Fig. 16) and ring gear (14) from pins (18). Be careful not to lose or damage shims (19). If puller is not available, it may be possible to use a punch through hole for plug (1) to carefully drive hub and ring gear from pins. Solid pins are used on later models in place of the three split pins (18). Remove pins only if necessary. Bump hub (22) and bearings (20 and 23) from steering knuckle (26). Remove cups for bearings (20 and 23) and seal (24) from hub. If necessary, remove inner bearing cone from knuckle housing.

Clean and inspect all parts for excessive wear or other damage and renew as necessary.

The steering knuckle and wheel hub can be assembled, then attached to the axle center member; however, the assembly is heavy and difficult to handle. End play of the steering knuckle bearings (32) must be measured and adjusted to provide 0.2 mm (0.008 inch) preload. End play measurement and adjustment is easier if the steering knuckle (26) is attached to the axle center member (27) before wheel hub and ring gear are assembled to the knuckle.

Be sure bushings (37 and 41) and seals (38 and 40) are installed, carefully locate the drive shaft and "U" joint assembly (39) in axle center member, then slide steering knuckle (26) over outer end of axle shaft. Lubricate "O" rings (31) and install in grooves of pivot pins (33) then press pins through steering knuckle bores and into bores in axle center member. Caps (29 and 35) and retaining screws (36) can be used to press pins (33) into bores if opposing screws (36) are carefully and evenly tightened. Be sure pins are fully seated in axle center member, remove caps (29 and 35) and install inner cones for both bearings (32) on pins. Install outer race for the lower bearing, install lower bearing cap and tighten the four retaining screws (36) to 96-118 N·m (71-87 ft.-lbs.) torque. Install the outer race for upper bearing, install upper cap (35) without shims (30) and tighten retaining screws to 96-118 N·m (71-87 ft.-lbs.) torque. Use a dial indicator and measure end play of the steering knuckle (26). Be sure that pins (33) and bearings (32) are fully seated. Without any shims (30), steering knuckle should have some end play. Remove upper cap (35), install shims (30) equal to the measured end play plus the desired bearing preload of 0.2 mm (0.008 inch), then reinstall upper cap (35). Tighten cap retaining screws to 96-118 N·m (71-87 ft.-lbs.) torque. Shims (30) can normally be added to the upper cap (35), but if a large number of shims are necessary, shims may be divided between the upper and lower caps to center the bearings (41 and 37) for axle (39).

Observe the following when reassembling wheel hub (22) to the steering knuckle (26). Drive cups for bearings (20 and 23) and oil seal (24) into hub (22). Lubricate bearings, bushings and seals liberally. If removed, install new cone for inner bearing (23) on knuckle housing (26). Install hub (22) and cone for outer bearing (20) onto knuckle housing. It may be necessary to heat bearing cones before installing. Install two screws with washers in opposing holes as shown in Fig. 18 to hold bearing (20—Fig. 16) in position while measuring standout of bearing race. Be sure that bearings are fully seated, then measure distance from end of steering knuckle hub and outer surface of outer bearing inner race. Measure standout of ring gear hub from the bearing contact surface to the steering knuckle hub contacting surface as shown in Fig. 19. Subtract measurement of ring gear hub (Fig. 19) from the bearing standout (Fig. 18), then add the desired bearing preload, 0.05 mm (0.002 inch) to determine the correct thickness of shims to install.

An example is:

Measurement shown in Fig. 18	8.10 mm	0.319 inch
Measurement shown in Fig. 19	− 6.20 mm	− 0.244 inch
	1.90 mm	0.075 inch
Desired bearing preload	+ 0.05 mm	+ 0.002 inch
Thickness of shims to install	1.95 mm	0.077 inch

Please note that measurements above are only examples to indicate the method of determining the correct thickness of shims to install. Selecting and installing the proper thickness of shims (19—Fig. 16) for each combination of parts will provide bearings

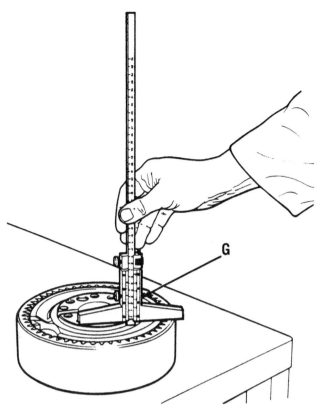

Fig. 19—Measure standout of ring gear hub as shown when selecting shims (19—Fig. 16). Refer to text.

(20 and 23) with correct, 0.05 mm (0.002 inch), preload.

After selecting the correct thickness of shims (19), install pins (18) if removed. If pins (18) are split type, install largest pin first, followed by medium size with gap located 180° from gap of the largest pin. Install smallest of the three pins with gap 180° from the middle size pin or nearly aligned with groove of the largest pin. On all models, remove screws (S—Fig. 18) and washers used while measuring. Install wheel hub (22), steering knuckle (26), ring gear (14), hub (17), shims (19) and related parts. Coat threads of screws (15) with "Loctite 270" and tighten evenly to 112-123 N.m (83-91 ft.-lbs.) torque.

Make sure thrust plate (5) is in place. Clean mating surfaces of wheel hub (22) and planet carrier (3), then coat face with "Loctite 515 Instant Gasket" and carrier assembly. Coat screws (2) lightly with oil before installing, then tighten to 96-118 N.m (71-87 ft.-lbs.) torque. Nut retaining the ball-joint of track rod end in the steering arm should be tightened to 78-86 N.m (58-63 ft.-lbs.) torque for 362-390T models, 108-118 N.m (80-87 ft.-lbs.) torque for 398 models. Nut retaining the ball-joint of steering cylinder end in the steering arm should be tightened to 98-108 N.m (72-80 ft.-lbs.) torque for 362-390T models, 177-196 N.m (130-145 ft.-lbs.) torque for 398 models.

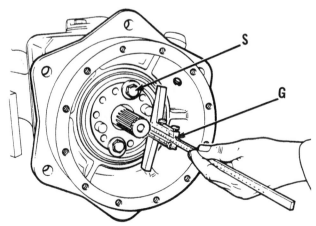

Fig. 18—Measure standout of inner bearing race as shown when selecting shims (19—Fig. 16). Refer to text.

Turn wheel hub (22) until hole for plug (1) is horizontal, fill hub and planetary to the level of opening for plug (1) with an approved oil, then install and tighten plug. Recommended oil for use in front drive axle is Massey-Ferguson Super 500 multi-use 10W-30 oil, Massey-Ferguson Permatran or equivalent. Install front wheel and tighten disc to hub nuts to 270 N·m (200 ft.-lbs.) torque. Disc to wheel rim nuts should be tightened to 190 N·m (140 ft.-lbs.) torque.

DIFFERENTIAL

All Except Models With "Hydralock"

13. REMOVE AND REINSTALL. The pinion shaft oil seal can be removed and a new seal installed, without removing the complete differential assembly. Mark the position of the pinion nut before removing, so that nut can be reinstalled without disturbing pinion bearing preload. Other service to the differential assembly requires removal of the unit.

To remove the differential assembly from front axle, refer to paragraph 9 and remove the drive shaft, then paragraph 12 and remove steering knuckles and axle shaft assemblies (39—Fig. 16) from both sides. Unbolt and remove the differential housing from the axle center housing. Some mechanics prefer to remove the front axle from tractor as outlined in paragraph 10 before removing steering knuckles and axle shafts.

Before reinstalling differential, clean mating surface of differential housing (1—Fig. 21) and axle center housing (27—Fig. 16), then coat mating surface with "Loctite 515 Instant Gasket." Lubricate

screws attaching differential to axle housing, install differential assemble and tighten retaining screws to 96-118 N·m (71-87 ft.-lbs.) torque. Install drive shaft coupling on pinion shaft, coat threads of center retaining screw with "Loctite 270," then tighten center screw to 60 N·m (44 ft.-lbs.) torque. Complete assembly by reversing the removal procedure. Refer to paragraph 9 for installing drive shaft.

14. OVERHAUL. Before disassembling, mark both bearing caps and housing as shown at (M—Fig. 20) to facilitate alignment when reassembling. Straighten tabs of lock plate (13—Fig. 21), then remove cap screw (12) and lock plate (13). Loosen, but do not remove, the four screws attaching caps (C), then unscrew adjusting rings (14). Remove both bearing caps and lift differential, bearing cups and adjusting rings from housing.

> NOTE: Some models are equipped with threaded adjusting rings (14—Fig. 20 and Fig. 21) on both sides, while others are equipped with spacer and shims (24 and 25—Fig. 22) on one side.

To remove the bevel pinion (10—Fig. 21), unstake nut (2), then remove nut and spacer (3). Push pinion

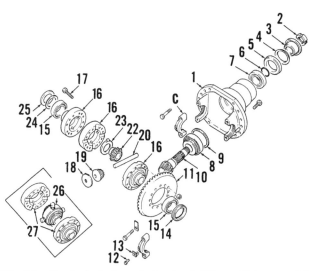

Fig. 21—Exploded view of standard differential assembly typical of the type used on many models. Gears (10 and 11) are available only as matched sets. "Autolock" differential unit (26) is shown in Fig. 23.

1. Housing	14. Adjusting ring
2. Nut	15. Bearing cup & cone
3. Spacer	16. Differential case
4. Seal	17. Screws
5. Seal	18. Thrust washers
6. "O" ring	19. Pinion gears
7. Bearing cup & cone	20. Pinion shaft
8. Bearing cup & cone	22. Side gears
9. Shim	23. Thrust washer
10. Bevel pinion	24. Shims
11. Ring gear	25. Spacer
12. Screw	26. "Autolock" differential
13. Locking clip	27. "Autolock" differential case

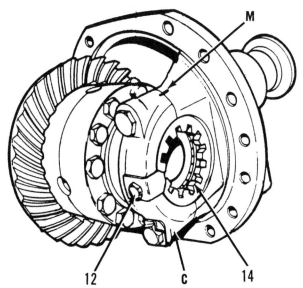

Fig. 20—View of removed differential assembly, showing marks (M) on one of the caps (C) and housing. Mark the other cap so that caps can be quickly identified for assembly to the correct side and in the correct position.

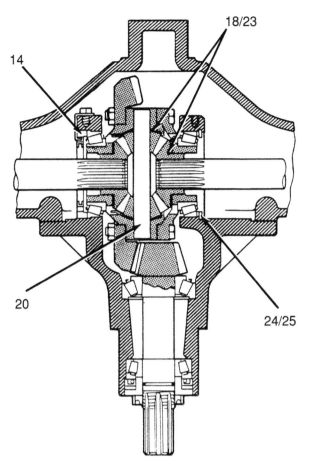

Fig. 22—Cross sectional view of differential of the type shown in Fig. 21. Refer to Fig. 21 for legend.

out toward inside. Cone of bearing (7) will slide from shaft as pinion is removed.

Mark differential case (16 or 27) and ring gear (11) to facilitate alignment when reassembling. On models with "Autolock" differential, refer to Fig. 24 and install washers of proper size and through-bolt as shown to hold the differential together. On all models,

remove screws (17—Fig. 21), then remove ring gear (11) and separate case (16). Note that two of the screws (17) are close fitting dowel screws and must be installed in same location when reassembling. Individual parts of the "Autolock" differential (Fig. 23) are not available and parts must be assembled in original location. Regardless of type, assembly of original parts is easier if parts are kept together for each side and not mixed.

Clean and inspect all parts for wear or other damage. Lubricate all parts while assembling. Align previously affixed marks on case (16 or 27—Fig. 21) and ring gear (11) while assembling. Coat threads of screws (17) with "Loctite 270," install and tighten screws to 79-87 N•m (58-64 ft.-lbs.) torque. Press bearing cones (15) onto case (16) until seated. Remove through-bolt and washers (Fig. 24) from "Autolock" models.

On all models, thickness of shims (9—Fig. 21) should be selected to change mesh position of bevel pinion (10) if gear set (10 and 11), pinion bearings (7 and 8) and/or differential housing (1) is renewed or if previous mesh position is questioned. The number stamped on the end of the pinion gear is the deviation from the standard distance of 118 mm or 113 mm (early axles with two adjuster rings for differential). Determine machined (cone) position by adding the stamped number (if marked with "+") or subtracting the stamped number (if marked with "–") from the standard distance. Install cup for bearing (8) in housing (1), then position cone for bearing (8) in cup as shown in Fig. 25. Measure distance (A—Fig. 25) from center line of the differential, as measured across the bearing saddle, to the face of inner race for pinion bearing (8). Subtract the measured distance (A—Fig. 25) from the machined cone distance (determined by adding or subtracting the number stamped on end of pinion to or from the standard dimension). The result is the thickness of shims (9—Fig. 21) to add under the bearing cup (8).

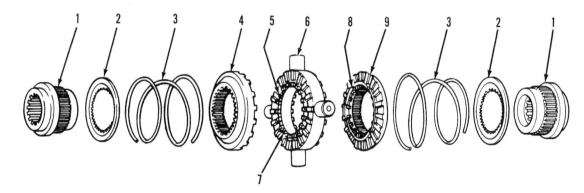

Fig. 23—Exploded view of "Autolock" differential. Washers and through-bolt shown in Fig. 24 must be used to compress springs (3) and hold assembly together while removing and installing.

1. Side gears
2. Spring retainers
3. Springs
4. Driven clutch assy.
5. Center cam
6. Central driver & center cam assy.
7. Snap ring
8. Holdout rings
9. Driven clutch

An example is:

Standard measurement . . . . 118.00 mm
Dimension stamped on pinion . . . . . + 0.20
                                       ─────────
Machined distance . . . . 118.20 mm
Measurement (A—Fig. 25) . . . . 119.50 mm
Machined distance . . . .   − 118.20 mm
                                       ─────────
Required shim thickness . . . . 1.30 mm

Please note that measurements above are only examples to indicate the method of determining the correct thickness of shims to install. Selecting and installing the proper thickness of shims (9—Fig. 21) for each combination of parts will provide gear set with proper mesh position. Also, note standard measurement (cone point) of some models is 113 mm.

Remove cup for bearing (8), install correct thickness of shims (9), then press cups for both bearings (7 and 8) into housing bores. Shims are available in various thicknesses from 0.10 mm to 1.00 mm. Install pinion (10), cone for bearing (7), "O" ring (6), seals (5 and 4) and spacer (3). Lubricate all parts liberally while assembling. Install a new special nut (2), tightening nut until 2.0-2.5 N•m (18-22 in.-lbs.) torque is required to turn pinion in the bearings. When correct rotational force is obtained, stake nut to shaft spline.

Reinstall differential and ring gear in housing (1—Fig. 21) by reversing removal procedure. Adjust gear backlash and carrier bearings as follows.

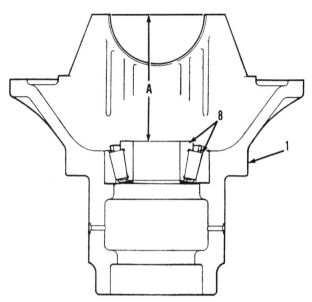

Fig. 25—Measure distance (A) from center line of differential, across bearing saddle, to the face of inner race for bearing (8).

Models with one adjuster ring (14—Fig. 20 and Fig. 21) are fitted with shims (24—Fig. 21) located on right side between spacer (25) and right side carrier bearing. The shims (24) are used to adjust backlash between the ring gear (11) and pinion (10). When assembling, install shims originally installed, install bearing caps (C) and tighten screws retaining caps to 136-150 N•m (100-110 ft.-lbs.) torque. Tighten adjuster ring (14) on left side until all clearance is just removed from carrier bearings (15), then tighten adjuster ring (14) three additional notches to provide correct bearing preload and align notch so that lock plate (13) can be installed. Do not yet install lock plate. Mount a dial indicator as shown in Fig. 26 and measure backlash between ring gear teeth and pinion gear teeth. Correct backlash is 0.10-0.25 mm (0.004-0.010 inch) when measured as shown. Reduce thickness of shims (24—Fig. 21) if backlash is excessive; increase thickness of shims if backlash is too tight.

On models with two adjuster rings (14—Fig. 20 and Fig. 21), one on each side, the setting of adjuster rings adjusts backlash of the bevel gear teeth and preload of carrier bearings. When assembling, install bearing caps (C) and tighten screws retaining caps to 136-150 N•m (100-110 ft.-lbs.) torque. Tighten both adjuster rings (14) until all clearance is just removed from carrier bearings (15), but be sure that some backlash exists between gear teeth. Tighten adjuster ring (14) on right side three additional notches to provide correct bearing preload and align notches so that lock plates (13) can be installed. Do not yet install lock plate. Mount a dial indicator as shown in Fig. 26 and measure backlash between ring gear teeth and pinion gear teeth. Correct backlash is 0.10-0.25 mm (0.004-

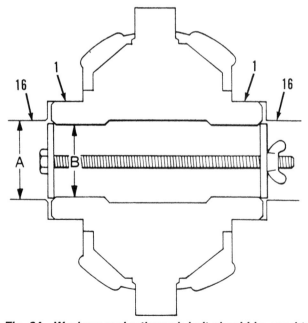

Fig. 24—Washers and a through-bolt should be used to hold "Autolock" differential assembly together while removing and installing. Diameter of washers should be small enough to pass through opening (A) in case (27—Fig. 21), but large enough to press against splines for axles in side gears (1—Fig. 23).

0.010 inch) when measured as shown. Loosen adjusting ring (14—Fig. 21) on left (ring gear) side, then turn adjuster on right side to change backlash. Readjust bearings by tightening adjuster ring on ring gear (left) side until all clearance is just removed from carrier bearings (15), then tighten left side adjuster ring three additional notches to provide correct bearing preload and align notches so that lock plates (13) can be installed.

On all models, recheck backlash after making any changes. With gear backlash, pinion bearing preload and carrier bearing preload correctly set, rotational torque of pinion shaft should be 3.2-3.3 N·m (28.32-29.21 in.-lbs.) torque. After it is determined that backlash and bearings are correctly set, carefully remove one of the four screws retaining caps (C), coat threads with "Loctite 270," then reinstall and tighten the screw to 136-150 N·m (100-110 ft.-lbs.) torque. Remove each of the remaining cap retaining screws one at a time, coat with "Loctite 270" and reinstall at correct torque. Coat threads of screws (12) with "Loctite 270," then install locking tabs (13) and retaining screws (12). Tighten screws (12) to 16-26 N·m (12-19 ft.-lbs.) torque.

Models With "Hydralock"

15. REMOVE AND REINSTALL. The pinion shaft oil seal can be removed and a new seal installed,

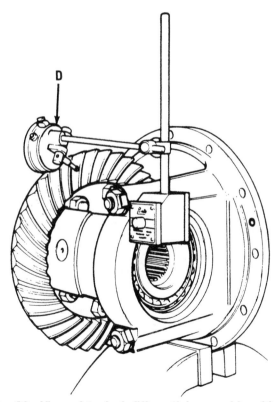

Fig. 26—View of typical differential assembly with dial indicator (D) mounted for checking gear backlash. Refer to text for procedure.

without removing the complete differential assembly. Mark the position of the pinion nut before removing, so that nut can be reinstalled without disturbing pinion bearing preload. Other service to the differential assembly requires removal of the unit.

To remove the differential assembly, refer to paragraph 9 and remove the drive shaft, then paragraph 12 and remove steering knuckles and axle shaft assemblies (39—Fig. 16) from both sides. Detach hydraulic hose and cover openings to prevent entrance of dirt. Unbolt and remove the differential housing from the axle center housing. Some mechanics prefer to remove the front axle from tractor as outlined in paragraph 10 before removing steering knuckles and axle shafts.

Before reinstalling differential, clean mating surface of differential housing (1—Fig. 27) and axle center housing (27—Fig. 16), then coat mating surface with "Loctite 515 Instant Gasket." Lubricate screws attaching differential to axle housing, install differential assemble and tighten retaining screws to 96-118 N·m (71-87 ft.-lbs.) torque. Install drive shaft coupling on pinion shaft, coat threads of center retaining screw with "Loctite 270," then tighten center screw to 60 N·m (44 ft.-lbs.) torque. Complete assembly by reversing the removal procedure. Refer to paragraph 9 for installing drive shaft.

16. OVERHAUL. Before disassembling, mark both bearing caps and housing as shown at (M—Fig. 20) to facilitate alignment when reassembling. Straighten tabs of lock plate (13—Fig. 27), then remove cap screw (12) and lock plate (13). Loosen, but do not remove, the four screws attaching caps (C), then unscrew adjusting ring (14). Remove both bearing caps carefully noting that the cap nearest the locking assembly retains a spring under compression. The spacer (25), spring retaining plate (28) and spring (29) should remain in the housing. Lift differential and carrier bearing cones from housing. Keep cups for carrier bearings (15) with the respective cones.

Install Massey-Ferguson MF471 or equivalent spring compressor (SC—Fig. 28) to compress spring (29—Fig. 27), then remove shims (24) and spacer (25). Bump the "Hydralock" unit (Fig. 28) toward center and remove from housing. "O" rings (36 and 37) are located in housing.

To remove the bevel pinion (10—Fig. 27), unstake nut (2), then remove nut and spacer (3). Push pinion out toward inside. Cone of bearing (7) will slide from shaft as pinion is removed.

Mark differential case (16) and ring gear (11) to facilitate alignment when reassembling. Remove screws (17), then remove ring gear (11) and separate case (16). Assembly is easier if parts are kept together for each side and not mixed.

Clean and inspect all parts for wear or other damage. Lubricate all parts while assembling. Align previously affixed marks on case (16) and ring gear (11) while assembling. Coat threads of screws (17) with "Loctite 270," install and tighten screws to 79-87 N•m (58-64 ft.-lbs.) torque. Press bearing cones (15) onto case (16) until seated.

Thickness of shims (9—Fig. 27) should be selected to change mesh position of bevel pinion (10) if gear set (10 and 11), pinion bearings (7 and 8) and/or differential housing (1) is renewed or if previous mesh position is questioned. The number stamped on the end of the pinion gear is the deviation from the standard distance of 118 mm or 113 mm (early axles

with two adjuster rings for differential). Determine machined (cone) position by adding the stamped number (if marked with "+") or subtracting the stamped number (if marked with "–") from the standard distance. Install cup for bearing (8) in housing (1), then position cone for bearing (8) in cup as shown in Fig. 25. Measure distance (A—Fig. 25) from center line of the differential, as measured across the bearing saddle, to the face of inner race for pinion bearing (8). Subtract the measured distance (A) from the machined cone distance (determined by adding or subtracting the number stamped on end of pinion to or from the standard dimension). The result is the thickness of shims (9—Fig. 27) to add under the bearing cup (8).

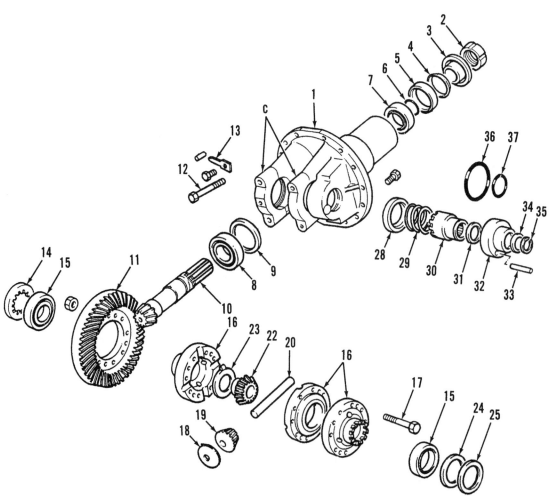

Fig. 27—Exploded view of "Hydralock" differential assembly. Gears (10 and 11) are available only as matched sets. Drawing shows assembly inverted, ring gear (11) is on opposite (left) side of pinion (10) as viewed when installed in tractor.

1. Housing	18. Thrust washers	29. Spring
2. Nut	19. Pinion gears	30. Sleeve
3. Spacer	20. Pinion shaft	31. Thrust washer
4. Seal	22. Side gears	32. Piston
5. Seal	23. Thrust washer	33. Dowel
6. "O" ring	24. Shims	34. Thrust washer
7. Bearing cup & cone	25. Spacer	35. Snap ring
8. Bearing cup & cone	28. Spring guide plate	36. "O" ring
9. Shim		37. "O" ring
10. Bevel pinion		
11. Ring gear		
12. Screw		
13. Locking clip		
14. Adjusting ring		
15. Bearing cup & cone		
16. Differential case		
17. Screws		

An example is:

Standard measurement	118.00 mm
Dimension stamped on pinion	+ 0.20
Machined distance	118.20 mm
Measurement (A—Fig. 25)	119.50 mm
Machined distance	− 118.20 mm
Required shim thickness	1.30 mm

Please note that measurements above are only examples to indicate the method of determining the correct thickness of shims to install. Selecting and installing the proper thickness of shims (9—Fig. 27) for each combination of parts will provide gear set with proper mesh position. Also, note standard measurement (cone point) of some models is 113 mm.

Remove cup for bearing (8), install correct thickness of shims (9), then press cups for both bearings (7 and 8) into housing bores. Shims are available in various thicknesses from 0.10 mm to 1.00 mm. Install pinion (10), cone for bearing (7), "O" ring (6), seals (5 and 4) and spacer (3). Lubricate all parts liberally while assembling. Install special nut (2), tightening nut until 2.0-2.5 N·m (18-22 in.-lbs.) torque is required to turn pinion in the bearings. When correct rotational force is obtained, stake nut to shaft spline.

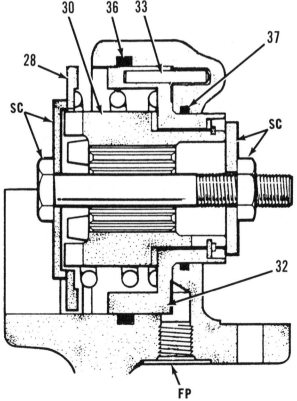

Fig. 28—Cross section of "Hydralock" unit with spring compressor (SC) in position. Fluid port is identified by (FP). Refer to Fig. 27 for remainder of legend.

Install and lubricate "O" rings (36 and 37—Fig. 27) in housing (1), then slide piston (32) into position making sure that pin (33) is in correct location. Install sleeve (30), thrust washers (31 and 34) and snap ring (35). Position spring (29) and retainer (28), then install spring compressor (SC—Fig. 28) so that spacer (25—Fig. 27) can be installed in housing groove.

Install spacer (25) and originally installed shims (24), then remove the spring compressor (SC—Fig. 28). Reinstall differential and ring gear in housing, then adjust gear backlash and carrier bearings as follows.

One adjuster ring (14—Fig. 27) is fitted, and shims (24) are located on right side between spacer (25) and right side carrier bearing. The shims (24) are used to adjust backlash between the ring gear (11) and pinion (10). When assembling, install shims originally installed, install bearing caps (C) and tighten screws retaining caps lightly. Use a pry bar to move the differential against the pressure of spring (29) and to remove all play in bearings (15), while tightening adjuster ring (14) to just remove all play. Tighten adjuster ring (14) three additional notches to provide correct bearing preload and align notch so that lock plate (13) can be installed after all clearance is just removed from carrier bearings (15). Do not yet install lock plate. Mount a dial indicator as shown in Fig. 26 and measure backlash between ring gear teeth and pinion gear teeth. Correct backlash is 0.10-0.25 mm (0.004-0.010 inch) when measured as shown. Reduce thickness of shims (24—Fig. 27) if backlash is excessive; increase thickness of shims if backlash is too tight. Recheck backlash after making any changes. After determining that backlash and bearings are correctly set, carefully remove one of the four screws retaining caps (C), coat threads with "Loctite 270," then reinstall and tighten the screw to 136-150 N·m (100-110 ft.-lbs.) torque. Remove each of the remaining cap retaining screws one at a time, coat with "Loctite 270" and reinstall at correct torque. Coat threads of screw (12) with "Loctite 270," then install locking tab (13) and retaining screw (12). Tighten screw (12) to 16-26 N·m (12-19 ft.-lbs.) torque.

TRACTOR SPLIT

Four-Wheel-Drive Models Without Cab

17. To separate tractors (without cab) between the rear of the transfer gearbox and the front of the rear axle center housing, proceed as follows:

Set parking brake and block the rear wheels to prevent rolling. Disconnect battery ground cable and drain all transmission fluid. Refer to paragraph 9 and remove the front drive shaft. Remove lower (gearbox) cover, both foot steps and battery boxes. Detach foot throttle, four-wheel-drive indicator switch and four-wheel-drive selector lever. Disconnect hydraulic oil

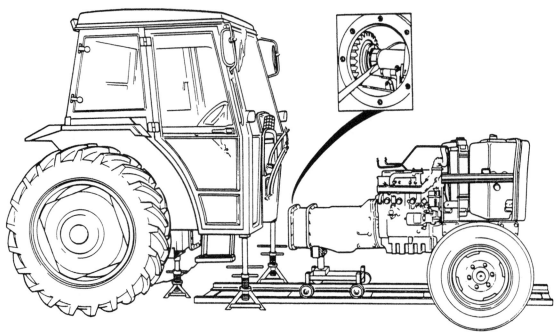

Fig. 29—View of tractor with cab separated between the four-wheel-drive transfer case and the rear axle center housing. Special splitting stand and track is shown.

supply pipe located on right side of tractor and, if so equipped, detach hydraulic pipes from left side. Disconnect brake lines and drain system. Disconnect wiring harness at the connectors below console on right side and disconnect wires from safety start switch. Remove the hydraulic suction filter assembly. Place hardwood wedges between the front axle and axle support casting on both side to prevent tipping around axle pivot when tractor is separated. Support rear of tractor under axle center housing and front of tractor under transmission housing. Remove screws attaching transfer case to the rear axle center housing, then separate the tractor halves. Either the front can be moved forward or the rear can be moved to the rear, but it is important to support both halves safely and securely.

If the split pin is removed from the shear tube, install pin in center of the five holes. Install shear tube on pinion. Install at least two dowel studs to facilitate alignment, position new gasket on dowel studs and position rear drive shaft into high/low gear coupler. Move the high/low shifter to "low" and shift main transmission to 3rd. Move tractor together, turning engine to align splines of shear tube. Make sure that flanges are tight against each other before installing retaining screws. Install and tighten one screw on each side, then check end float of the rear drive shaft. Push the shear tube toward rear as shown at inset in Fig. 29, which will compress spring, then measure clearance between the shear tube and locking collar. Correct clearance is 0.4-2.5 mm (0.015-

0.100 inch) and can be changed by separating tractor and moving the split ring to another hole in the shear tube. Tighten all screws retaining the transfer case to the rear axle center housing to 112 N·m (83 ft.-lbs.) torque, beginning at top center and progressing in a clockwise direction (as viewed from tractor rear) two times around the mating flange. Remainder of assembly is reverse of disassembly.

Four-Wheel-Drive Models With Cab

18. To separate tractors (with cab) between the rear of the transfer gearbox and the front of the rear axle center housing, proceed as follows:

Set parking brake, block the rear wheels to prevent rolling and drain all transmission fluid. Disconnect battery ground cable, then disconnect wiring harness at the connectors. Refer to paragraph 9 and remove the front drive shaft. Disconnect throttle stop cable and throttle cable from fuel injection pump and move cables out of the way so that cables will not catch when separating tractor. Disconnect and cap all hydraulic pipes running from front to rear, which would prevent separation of tractor halves. Mark (identify), disconnect and cap steering hoses, heater hoses and air conditioning lines (if so equipped), which would interfere. Place hardwood wedges between the front axle and axle support casting on both sides to prevent tipping around axle pivot when tractor is separated. Remove front floor mat and floor plates, disconnect clutch linkage and detach gearbox oil filler pipe.

Unbolt and remove the gear shift cover complete with shift levers, then unbolt and remove the transmission cover. Disconnect wires at safety start switch and four-wheel-drive indicator switch. Remove the oil feed pipe and filter housing from the right side of transmission case. Disconnect four-wheel-drive selector lever, then check to make sure that all wiring, cables, linkages and hoses are free to permit the tractor to be separated. Support rear of tractor under axle center housing and at front corners of cab. Support front of tractor under transmission housing in such a way that front of tractor, engine, transmission and transfer case can be rolled forward (Fig. 29). Unbolt and remove both cab support brackets from front of cab. Remove screws attaching transfer case to the rear axle center housing, then move front of tractor forward. It is important to support both halves safely and securely to prevent damage and injury.

If the split pin is removed from the shear tube, install pin in center of the five holes. Install shear tube on pinion. Install at least two dowel studs to facilitate alignment, position new gasket on dowel studs and position rear drive shaft into high/low gear coupler. Move the high/low shifter to Low and shift main transmission to 3rd. Move front half of tractor toward rear turning engine to align splines of shear tube. Make sure that flanges are tight against each other before installing retaining screws. Install and tighten one screw on each side, then check end float of the rear drive shaft. Push the shear tube toward rear as shown at inset in Fig. 29, which will compress spring, then measure clearance between the shear tube and locking collar. Correct clearance is 0.4-2.5 mm (0.015-0.100 inch) and can be changed by separating tractor and moving the split ring to another hole in the shear tube. Tighten all screws retaining the transfer case to the rear axle center housing to 112 N·m (83 ft.-lbs.) torque, beginning at top center and progressing in a clockwise direction (as viewed from tractor rear) two times around the mating flange. Remainder of assembly is reverse of disassembly.

TRANSFER GEARBOX

Power for the front drive axle is taken from a gearcase and clutch (Fig. 35) attached to the left side of the range gearbox of models with the twelve-speed shuttle transmission. Power for the front axle of models equipped with other transmissions, is taken from a transfer gearbox (Fig. 30 or Fig. 32) located behind the tractor standard transmission. Refer to the appropriate following paragraphs for service.

Four-Wheel-Drive Models With All Transmissions Except Twelve-Speed Shuttle

19. R&R AND OVERHAUL. To remove the transfer gearbox, first split the tractor between the rear of the transfer gearbox and the front of the rear axle center housing as outlined in paragraph 17 or 18, then proceed as follows.

Position floor jack under the transfer gearbox, remove retaining screws and lower transfer gearcase from the tractor.

To disassemble the unit, refer to Fig. 30 and Fig. 31 or Fig. 32 and Fig. 33 and proceed as follows. Remove input gear (2), then unbolt and remove retainer (5). Drive input shaft (3) and bearing (7) from housing. Remove snap ring (16) from shaft (11), then unbolt and remove retainer (10). Drive gear and shaft (11) out toward rear and withdraw gear (13) from housing. Remove snap ring (24), then use a 12 mm screw in threaded end of shaft (21) to withdraw shaft from housing bore. Lift gear (19), bearings (18) and thrust washers (17 and 20) from housing. Remove the center screw from drive shaft coupling, then remove coupling and drive shaft guard.

On early models (before July 1991), unbolt and remove cover (40—Fig. 30). Remove screw (44) from fork (43), then remove shaft (46), shift fork (43) and shoes (42). Remove output shaft (37) and related parts. Be careful not to lose the three sets of detent balls and springs (38).

On later models (from July 1991), unbolt and remove cover (61—Fig. 32), then remove screw (60) from fork. Pull shaft (63) from housing (57) and remove fork (59). Unbolt housing (57) and withdraw output shaft (37) and related parts. Be careful not to lose the two detent balls and one spring (55) if actuator (54) is removed from shaft (37).

Clean and inspect all parts and renew any showing excessive wear or other damage. When reassembling, renew all "O" rings, seals and gaskets. Reassemble by reversing disassembly procedure.

Reinstall transfer gearbox by reversing the removal procedure. Tighten transfer gearbox retaining screws to 112 N·m (83 ft.-lbs.) torque, beginning at top center and progressing in a clockwise direction (as viewed from tractor rear) two times around the mating flange. Coat threads of drive shaft retaining screws with "Loctite 270," then attach drive shaft flanges and tighten screws to 55-75 N·m (40-55 ft.-lbs.) torque. Remainder of assembly is reverse of disassembly.

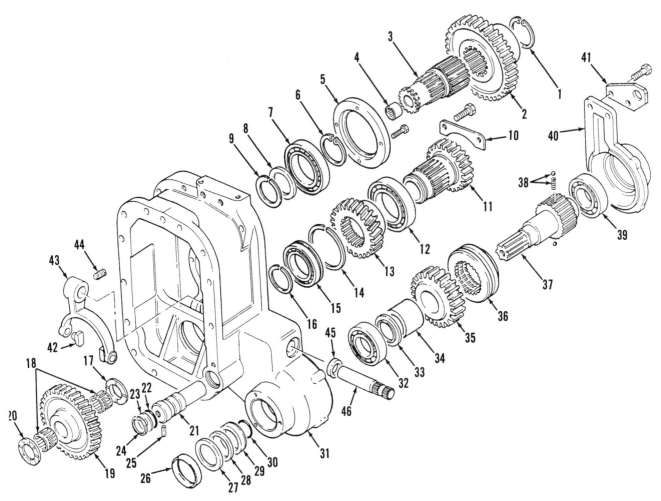

Fig. 30—Exploded view of four-wheel-drive transfer gearbox used on early models (before July 1991) without twelve-speed shuttle transmission. Refer to Fig. 32 for later type.

1. Snap ring
2. Input gear
3. Input shaft
4. Needle bearing
5. Bearing retainer
6. Snap ring
7. Ball bearing
8. Washer (365 models)
9. Snap ring
10. Retainer plate
11. Gear and shaft
12. Roller bearing
13. Gear
14. Snap ring
15. Ball bearing
16. Snap ring
17. Thrust washer
18. Roller bearings (2)
19. Intermediate gear
20. Thrust washer
21. Intermediate shaft
22. "O" ring
23. Shim (0.1-0.2 mm)
24. Snap ring
25. Pin
26. Sleeve
27. Retainer
28. Seal
29. Seal
30. "O" ring
31. Transfer gearbox
    housing
32. Ball bearing
33. Spacer
34. Bushing
35. Output gear
36. Engagement coupling
37. Output shaft
38. Detent balls and springs (3)
39. Ball bearing
40. Cover
41. Bracket
42. Shoes (2)
43. Shift fork
44. Set screw
45. Seal
46. Shift shaft

Four-Wheel-Drive Models With Twelve-Speed Shuttle Transmission

20. R&R AND OVERHAUL. To remove the four-wheel-drive transfer gearbox for these models, first drain oil from the transmission and detach rear of drive shaft from the coupling (1—Fig. 34). Disconnect hydraulic supply line (2) and wires (3). Support the gearbox, remove the retaining screws, then remove the unit from side of transmission case.

To disassemble the removed unit, first unscrew the solenoid valve (33—Fig. 35), remove the center screw (1) and remove the coupling (3). Remove screws (35) and cap (36), being careful not to lose shims (37), then push shaft (38) out of intermediate gear (41) and housing. Lift intermediate gear, bearings (39), snap rings (40) and thrust washer (42) from housing. Remove screws retaining cover (32), then remove cover, output shaft (20) and associated parts. Unbolt and remove cover (4), end cap (7), seals and associated parts being careful not to lose or damage shims (9).

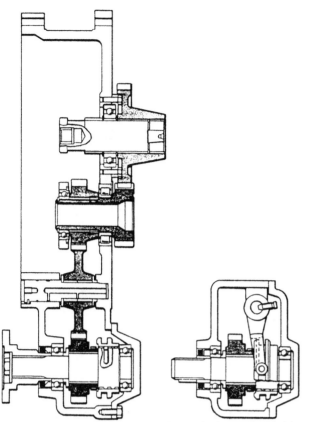

Fig. 31—Cross section of early style four-wheel-drive transfer gearbox shown exploded in Fig. 30.

Fig. 32—Exploded view of late style (from July 1991) four-wheel-drive transfer gearbox used on tractors without twelve-speed shuttle transmission.

1. Snap ring
2. Input gear
3. Input shaft
4. Roller bearing
5. Bearing retainer
6. Snap ring
7. Ball bearing
8. Washer (365 models)
9. Snap ring
10. Retainer plate
11. Gear and shaft
12. Roller bearing
13. Gear
14. Snap ring
15. Ball bearing
16. Snap ring
17. Thrust washer
18. Roller bearings (2)
19. Intermediate gear
20. Thrust washer
21. Intermediate shaft
22. "O" ring
23. Shim (0.1-0.2 mm)
24. Snap ring
25. Pin
26. Sleeve
27. Seal
29. Seal
30. "O" ring
31. Transfer gearbox housing
32. Ball bearing
33. Thrust washer
35. Output gear
37. Output shaft
47. Snap ring
48. Spring
49. Engagement coupling
50. Toggle (engagement) pins (3)
51. Abutment ring
52. Snap ring
53. Ball bearing
54. Actuator shaft
55. Detent spring and 2 balls
56. "O" ring
57. Shift housing
58. Pins (2)
59. Shift fork
60. Set screw
61. Cover
62. Seal
63. Shift shaft

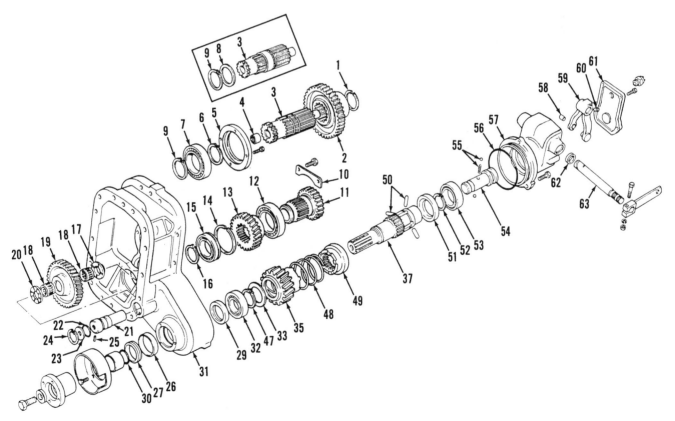

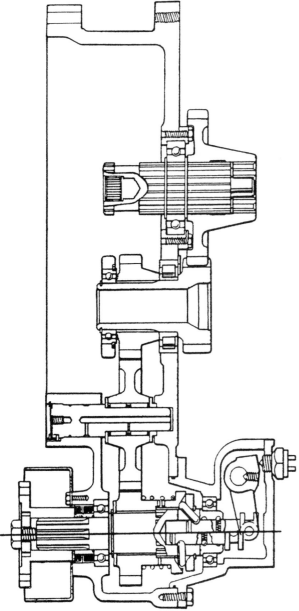

Fig. 33—Cross section of late style four-wheel-drive transfer gearbox shown exploded in Fig. 32.

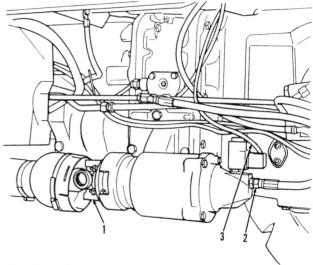

Fig. 34—View of four-wheel-drive transfer gearbox mounted on left side of tractor with twelve-speed shuttle transmission.

1. Driveshaft rear coupling
2. Wires to engagement solenoid
3. Hydraulic pressure line

Press bearing cone (13) from shaft (20), then remove thrust washer (14) and gear (15). A steel tube with outside diameter of approximately 16 mm (⅝ inch), inside diameter of approximately 13 mm (½ inch) and approximately 100 mm (4 inches) long is necessary to remove and disassemble actuator piston (23). Use special tube and a press to push against washer (27) and compress spring (26) enough to remove snap ring (28). Remainder of disassembly will be self-evident. Refer also to cross section shown in Fig. 36.

Clean and inspect all parts and renew any showing excessive wear or other damage. When reassembling, renew all "O" rings, seals and gaskets. Reassemble as

follows. Install gear (15—Fig. 35) and thrust washer (14) on output shaft (20), then press bearing cone (13) onto shaft against thrust washer and shoulder on shaft. Assemble spring (16), coupler (17), abutment ring (18) and snap ring (19) on shaft, then press bearing cone (29) onto output shaft (20). Install "O" rings (22 and 24) in grooves of actuator piston (23), lubricate piston and insert into bore of output shaft (20). Install toggle pins (21) between abutment ring (18) and coupler (17), making sure that pins enter the groove of actuator piston (23), then slide snap ring (19) into groove of output shaft (20). Install spring (26), sleeve (25) and washer (27), compress spring sufficiently so that snap ring (28) can be installed in groove of output shaft. Measure and adjust preload of bearings (13 and 29) before installing new seals (5 and 10). Assemble shaft assembly (13 through 29), rear cover (32) and front cover (7) using shims (9) originally installed. Measure bearing preload and vary the thickness of shims (9) as required to establish correct preload of 0.07 mm (0.003 inch). When preload is correct, a slight resistance should be felt when rotating the shaft by hand. There must be no free play in the bearings. Remove output shaft assembly to install intermediate gear and shaft assembly (38 through 42), measure preload of bearings (39) and select correct thickness of shims (37). Install sufficient thickness of shims (37) to provide bearings (39) with 0-0.05 mm (0-0.002 inch) preload. When preload is correct, there should be slight resistance felt when rotating the gear by hand. Reinstall shaft and bearing assembly (13 through 29), using new "O" rings (8, 12 and 31) and seals (5 and 10). Remainder of reassembly is reverse of disassembly. Degrease threads for screw (1), coat threads with "Loctite 270," then tighten to 113-137 N•m (83-101 ft.-lbs.) torque. Degrease mating surface of housing (34) and transmis-

sion, coat mating surfaces with "Loctite 515 Instant Gasket" and tighten retaining screws to 120-160 N•m (88-118 ft.-lbs.) torque. Tighten drive shaft coupling screws to 55-75 N•m (40-55 ft.-lbs.) torque.

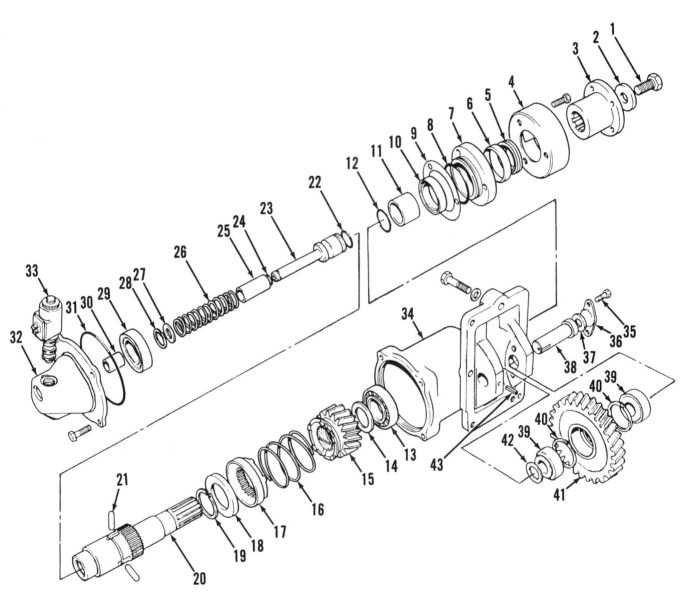

Fig. 35—Exploded view of transfer gearbox used on four-wheel-drive models with twelve-speed shuttle transmission.

1. Screw	12. "O" ring	23. Actuator piston	
2. Washer	13. Tapered roller bearing	24. "O" ring	34. Housing
3. Driveshaft coupling	14. Thrust washer	25. Stop tube	35. Screw
4. Cover	15. Drive gear	26. Spring	36. End cap
5. Seal	16. Spring	27. Washer	37. Shims
6. Sleeve	17. Clutch coupler	28. Snap ring	38. Intermediate shaft
7. End cap	18. Abutment ring	29. Tapered roller bearing	39. Tapered roller bearings
8. "O" ring	19. Snap ring	30. Insert	40. Snap rings
9. Shims	20. Output shaft	31. "O" ring	41. Intermediate gear
10. Oil seal	21. Toggle pins (3)	32. End cover	42. Thrust washer
11. Oil seal sleeve	22. "O" ring	33. Solenoid valve	43. Pin

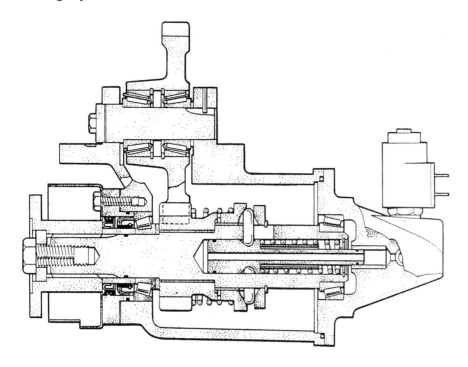

Fig. 36—Cross sectional view of four-wheel-drive unit shown exploded in Fig. 35.

STEERING SYSTEM

21. The hydrostatic steering system consists of a pump, steering valve assembly, and one or two steering cylinders. In normal operation, pressurized hydraulic fluid is supplied by the gear type hydraulic pump. However, in the event of hydraulic failure or engine stoppage, manual steering can be accomplished by the gerotor pump in the steering valve.

FILTER AND BLEEDING

All Models

22. Recommended steering fluid is Massey-Ferguson Super 500 multi-use 10W-30 oil or equivalent.

On models with a separate steering system pump and reservoir, maintain fluid level at bottom of reservoir filler hole with tractor level. Fluid and filter should be changed if steering unit is overhauled or if fluid is suspected of contamination. The tandem (dual) pump, used on some models, is supplied with oil from the gearbox and rear axle center housing. A spin-on filter is located near the pump to filter fluid returning from the steering control valve to the auxiliary section of the pump.

On models with separate pump, filter (3—Fig. 37) can be renewed as follows. Remove screw or nut (1) retaining reservoir (2), then pull reservoir from the pump unit and catch oil in a suitable container. Re-

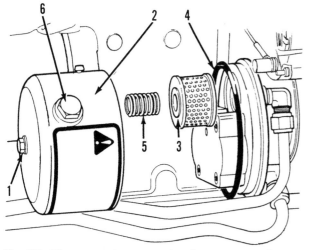
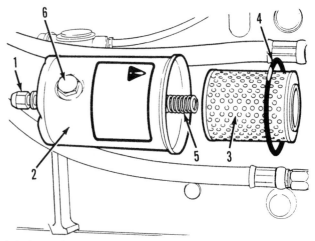

Fig. 37—The separate power steering pump used on some models is equipped with a filter as shown. Pump may be one of several different types, but reservoir and filter are serviced in a similar way.

move the filter assembly (3) and sealing ring (4). Clean the inside of the reservoir and outside of the pump. Install new sealing ring (4) and filter, then install reservoir, aligning breather at top before tightening screw or nut (1).

The hydrostatic steering system on all models is self-bleeding. When the unit has been disassembled, refill with new oil, then start engine and cycle the system several times by turning the steering wheel from lock to lock. Recheck fluid level and add fluid as required. Capacity of separate reservoir is approximately 1.2 L (2.1 pints). System is fully bled when no more air bubbles appear in reservoir as system is cycled.

TROUBLESHOOTING

All Models

23. Some of the problems which may occur during operation of power steering and their possible causes are as follows:
1. Steering wheel hard to turn. Could be caused by:
   a. Defective power steering pump.
   b. Leaking or missing recirculating ball.
   c. Mechanical parts of front steering system binding.
   d. Ball bearings in steering column damaged.
   e. Leaking steering cylinder.
   f. Control valve spool and sleeve binding.
2. Steering wheel turns on its own. Could be caused by:
   a. Leaf springs in control valve weak or broken.
3. Steering wheel does not return to neutral position. Could be caused by:
   a. Control valve spool and sleeve jammed.
   b. Leakage between valve sleeve and housing.
   c. Dirt or metal chips between valve spool and sleeve.
4. Excessive steering wheel play. Could be caused by:
   a. Inner teeth of rotor or drive shaft teeth worn.
   b. Upper flange of drive shaft worn.
   c. Leaf springs in control valve weak or broken.
5. Steering wheel rotates at steering cylinder stops. Could be caused by:
   a. Excessive leakage in steering cylinder.
   b. Rotor and stator excessively worn.
   c. Excessive leakage between valve spool and sleeve.
   d. Excessive leakage between sleeve and housing.
6. Steering wheel "kicks" violently. Could be caused by:
   a. Incorrect adjustment between drive shaft and rotor.
7. Steering wheel responds too slowly. Could be caused by:
   a. Excessive front end weight.
   b. Not enough oil.
   c. Steering control valve worn.

8. Tractor steers in wrong direction. Could be caused by:
   a. Hoses to steering cylinder incorrectly connected.
   b. Incorrect timing of drive shaft to rotor.

SYSTEM PRESSURE

Models With Separate Steering Pump

24. To check the system relief pressure, install a "T" fitting in pump pressure line at pump (Fig. 38), connect a 0-28,000 kPa (0-4000 psi) test gauge to the fitting and operate engine at 1200 rpm. Turn the front wheels to one extreme against lock, and observe gauge reading. System relief pressure should be 17,262 kPa (2466 psi) for models so equipped.

> CAUTION: When checking system relief pressure, hold the steering wheel against lock only long enough to observe pressure indicated by gauge. Pump may be damaged if steering wheel is held in this position too long or if flow is otherwise stopped.

The pressure relief valve is located in the steering control valve as shown at (34, 35 and 36—Fig. 49). Adjustment requires removal of the steering control valve as described in paragraph 33. Turning the relief valve spring seat (34) in, increases the pressure against spring (35) and increases system relief pressure. Spring seat plug (34) is accessible after removing plug (32).

Models With Tandem Pump

25. To check the system relief pressure, install a "T" fitting in pump pressure line at pump (Fig. 39), con-

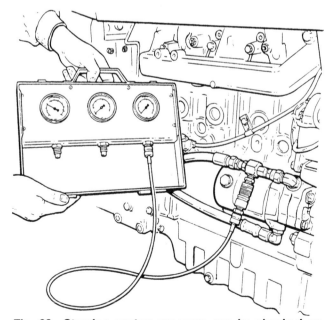

Fig. 38—Steering system pressure can be checked on systems with separate pump as described in text after attaching pressure gauge as shown.

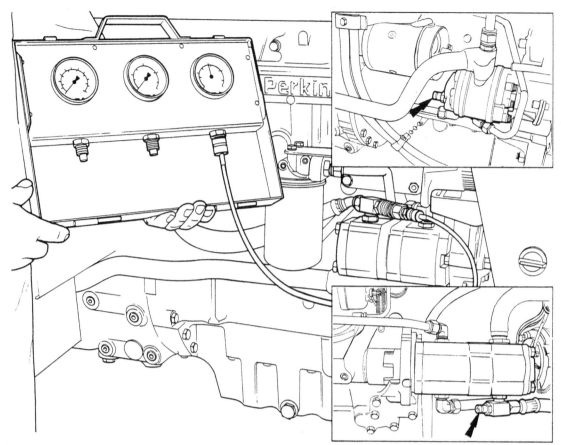

Fig. 39—Check steering system pressure with gauge attached as shown on models with tandem pump. System relief pressure is calculated by subtracting the circuit back pressure (pressure without turning wheels) from the higher pressure indicate on gauge with wheels held against one lock.

nect a 0-28,000 kPa (0-4000 psi) test gauge to the fitting and operate engine at 1200 rpm. Observe back pressure on gauge, then turn the front wheels to one extreme against lock, and observe gauge reading. Pressure when wheels are held against lock less the normal back pressure is the system relief pressure. System relief pressure should be 15,855-17,923 kPa (2300-2600 psi) for four cylinder models.

> **CAUTION: When checking system relief pressure, hold the steering wheel against lock only long enough to observe pressure indicated by gauge. Pump may be damaged if steering wheel is held in this position too long or if flow is otherwise stopped.**

The pressure relief valve is located in the steering control valve as shown at (34, 35 and 36—Fig. 49). Adjustment requires removal of the steering control valve as described in paragraph 33. Turning the relief valve spring seat (34) in, increases the pressure against spring (35) and increases system relief pressure. Spring seat plug (34) is accessible after removing plug (32).

PUMP

All Models

26. A tandem (dual section) pump is used on models equipped with spool valves and other auxiliary hydraulics. One section of the pump supplies pressure for the steering. Pressurized fluid not required for steering passes through a spin-on type filter. The second section of the pump supplies pressurized fluid to the auxiliary spool valves, Multi-Power transmission (if so equipped) and independent pto clutch (if so equipped). Refer to paragraphs 29 and 30 for service to tandem hydraulic pump.

A separate steering pump containing its own reservoir is used on some models requiring hydraulic pressure for steering only. Refer to paragraphs 27 and 28 for service to the separate pump used on these models.

Models With Separate Pump

27. REMOVE AND REINSTALL. Clean pump and area around pump thoroughly, then disconnect lines from pump and allow fluid to drain. Cover all

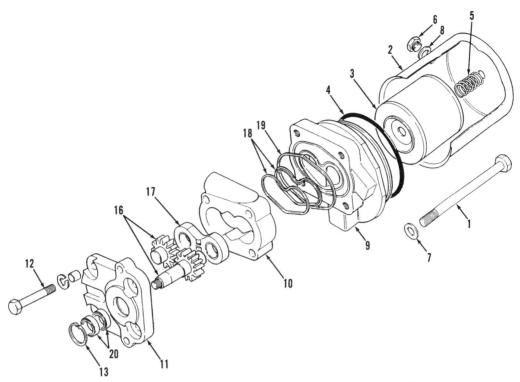

Fig. 40—Exploded view of separate steering pressure pump used on some models. Unit shown is made by Sundstrand Hydraulic.

1. Screw				16. Pump gears	
2. Reservoir	6. Fill plug	10. Pump body	17. Bearing		
3. Filter element	7. Gasket	11. Cover and flange	18. Inner seals		
4. "O" ring	8. Gasket	12. Screw	19. Outer seal		
5. Spring	9. Cover	13. Snap ring	20. Seals		

openings to prevent dirt from entering pump or lines, then unbolt and remove pump from the engine timing case.

Tighten pump retaining screws to 31-42 N·m (23-31 ft.-lbs.) torque. Refer to paragraph 22 for filling and bleeding.

28. OVERHAUL. Refer to Fig. 40 and straighten tab of lock washer. Remove nut, then use a puller to remove drive gear from pump shaft. Remove Woodruff key after gear is pulled from shaft. Remove screw or nut (1), then remove reservoir (2) and filter (3). Mark relative position of flange housing (11), body (10) and cover (9), then remove bolts (12). Pump gears (16) are available only as matched set. Check condition of bearing blocks (17), gears (16) and body (10) for wear or other damage.

Always use new filter and seals when assembling. Tighten pump body through-bolts (12) to 48-50 N·m (35-37 ft.-lbs.) torque and pump drive gear retaining nut to 40-45 N·m (30-33 ft.-lbs.) torque.

Models With Tandem Pump

29. REMOVE AND REINSTALL. Clean pump and area around pump thoroughly, then disconnect lines from pump and allow fluid to drain. Cover all openings to prevent dirt from entering pump or lines, then unbolt and remove pump from the engine timing case.

Tighten pump retaining screws to 31-42 N·m (23-31 ft.-lbs.) torque. Refer to paragraph 22 for filling and bleeding.

30. OVERHAUL. Several different pumps have been used on these models. Refer to the following and Fig. 41, Fig. 42 and Fig. 43 for typical service.

Straighten tab of lock washer, remove nut, then use a puller to remove the pump drive gear. Remove Woodruff key after gear is pulled from shaft. Mark relative position of flange housing (11), bodies (2 and 10) and covers (1, 3 and 9), then remove bolts (12). Pump gears (4 and 16) are available only as matched set. Check condition of bearing blocks (5 and 17), gears (4 and 16) and bodies (2 and 10) for wear or other damage.

Fig. 41—Exploded view of Bosch tandem pump used on some models.

1. Cover
2. Pump body
3. Cover
4. Pump gears
5. Bearings
6. Inner seals
7. Outer seal
8. Coupling
9. Cover
10. Pump body
11. Cover and flange
12. Screw
13. Snap ring
14. Seal
15. Woodruff key
16. Pump gears
17. Bearings
18. Inner seals
19. Outer seal

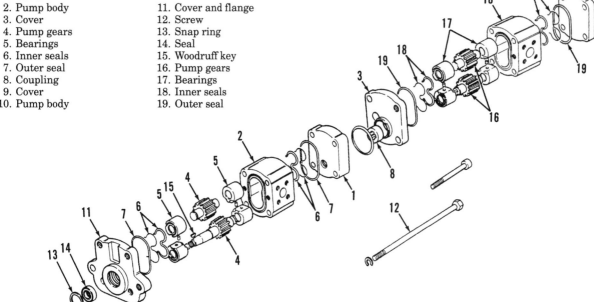

Fig. 42—Exploded view of Sundstrand tandem pump used on some models.

1. Pump center housing
2. Pump body
4. Pump gears
5. Bearings
6. Inner seals
7. Outer seals
8. Coupling
9. Cover
10. Pump body
11. Cover and flange
12. Screw
14. Seal
16. Pump gears
17. Bearings
18. Inner seals
19. Outer seals

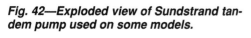

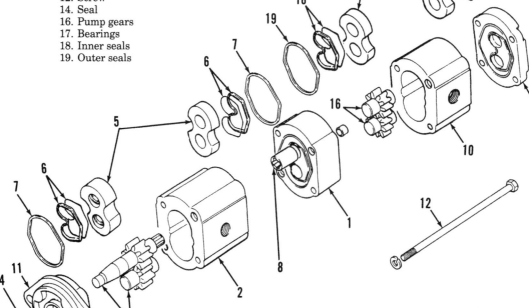

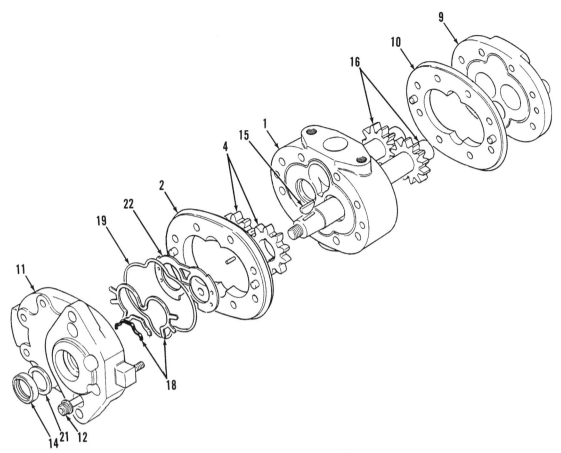

Fig. 43—Exploded view of Cessna tandem pump used on some models.

1. Pump center housing
2. Pump body
4. Pump gears
9. Cover
10. Pump body
11. Cover and flange
12. Screw
14. Seal
15. Woodruff key
16. Pump gears
18. Inner seals
19. Outer seals
21. Washer
22. Wear plate

On Bosch pump (Fig. 41), mark location of each part of bearings (5 and 17) to be sure parts are reinstalled in same location. Grooves less than 0.3 mm (0.012 inch) deep are permissible in the larger (inlet) hole of pump bodies (2 and 10). Check and install new inner seals (6 and 18) as follows: Height of seal support strip should be 1.95-2.10 mm (0.077-0.085 inch). If any part of support strip is too high, turn strip upside down in a cover and carefully polish lightly with emery paper. Install support strips and seal strips in covers with beveled edge of support strips and rounded edge of seal strips against rounded (radius) sides of seal groove. Marks affixed to parts before disassembling, should be aligned when reassembling. Support strip is toward suction side and seal strip is toward pressure side. Tighten M8 screws (12) to 17-23 N•m (13-17 ft.-lbs.) torque and M10 screws to 35-46 N•m (26-34 ft.-lbs.) torque. Tighten screws evenly and check for free rotation of pump shaft while tightening. Tighten pump drive gear retaining nut to 40-45 N•m (30-33 ft.-lbs.) torque.

On Sundstrand pumps (Fig. 42), bearing (5) nearest the cover with mounting flange (11) is thicker than other bearings and extends into front cover. Pack shaft seal (14) with grease before installing, and stick seals (6, 7, 18 and 19) into position with grease while assembling. Marks affixed to parts before disassembling, should be aligned when reassembling. Tighten screws (12) to 54-61 N•m (40-45 ft.-lbs.) torque. Tighten screws evenly and check for free rotation of pump shaft while tightening. Tighten pump drive gear retaining nut to 40-45 N•m (30-33 ft.-lbs.) torque.

On Cessna pumps (Fig. 43), auxiliary pump gears (4) should be at least 12.95 mm (0.510 inch) thick and steering pump gears (16) should be at least 9.75 mm (0.384 inch) thick. Shaft bushings are not available separately from covers (1, 9 and 11). If inside diameter of bushings exceeds 19.177 mm (0.755 inch), install new covers and bushing assemblies. Inside diameter of gear pockets in bodies (2 and 10) should be 43.51 mm (1.713 inch) or less. The two half moon cavities in pump bodies (2 and 10) must face away from front plate (11) and the smaller half moon cavity must be on pressure side of pump. Install wear plate (22) with bronze side toward gears and larger cut-away in middle section toward suction side of pump. Marks affixed to parts before disassembling should be aligned when reassembling. Install washer (21) and seal (14) after pump is assembled and screws (12) are installed and tightened to 37-40 N·m (27-30 ft.-lbs.) torque. Tighten screws evenly and check for free rotation of pump shaft while tightening. Lubricate seal (14), washer (21) and pump drive shaft. Install washer (21), position a 33 mm (1.3125 inches) diameter seal protector over shaft, then install seal (14) being careful not to damage seal lip. Press seal into front cover until seal is 0.5-1.0 mm (0.020-0.040 inch) below front face of cover (11). Tighten pump drive gear retaining nut to 40-45 N·m (30-33 ft.-lbs.) torque.

To run in new or rebuilt pump, install pump and run engine for three minutes at 1200 rpm at 0 (zero) hydraulic pressure, only circulating hydraulic fluid. Operate auxiliary control valve intermittently (to build pressure) for three minutes, then turn the steering wheel lock to lock for three minutes. Increase engine speed to rated rpm and recheck auxiliary and steering systems. Idle engine and check for leaks. Change filter and refill transmission oil.

POWER CYLINDER

Two-Wheel-Drive Models

31. R&R AND OVERHAUL. The cylinder is attached between axle and steering arms as shown in Fig. 44. To remove the cylinder, first refer to paragraph 4 and remove the axle and steering cylinder as an assembly. The steering cylinder is attached to the axle center member with pivot block (8—Fig. 8). Remove cotter pin, remove nut, then bump tapered end of ball-joint (3) from pivot block (8). Complete removal of steering cylinder by detaching rod ends from steering arms (1).

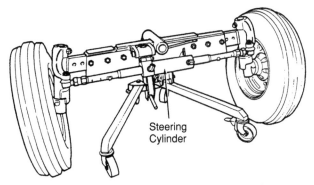

Fig. 44—Steering cylinder can be removed from two-wheel-drive tractors after first removing axle assembly.

Refer to Fig. 8 and remove the track rods (6) and rod ends. Ball-joint (3) can be removed after removing the clamp screw. Clamp the cylinder (1—Fig. 45) in a holding fixture and clean sealing compound from elongated hole (H). End cap (7) has two holes which can be used to rotate the end cap until the end of the retaining ring (2) is located in the hole. Use a small screwdriver or similar instrument to pull and guide the end of the retaining ring out through the elongated hole (H). Turn the end cap (7) and pull the retaining ring from between the end cap and cylinder. One end of the retaining ring is bent so that end enters a hole in the end cap. After retaining ring (2) is removed, withdraw piston and rod (12), cap (7) and seals from cylinder (1). Remove old seals and install new seals before reassembling.

Lubricate and assemble steering cylinder by reversing disassembly procedure. Turn end cap (7) in cylinder until the hole in end cap for end of the retaining ring (2) is visible in hole (H). Insert the bent end of retaining ring in hole and turn end cap, drawing retaining ring (2) into groove formed between end

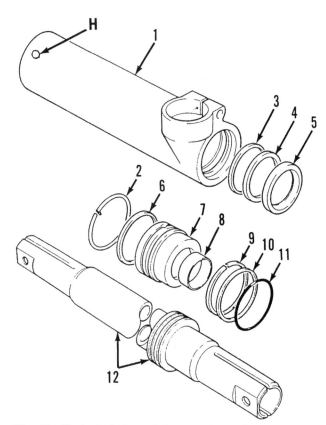

Fig. 45—Exploded view of the steering cylinder used on two-wheel-drive tractors. Sealing compound is used to seal hole (H) after assembly and installation of retaining ring (2).

1. Steering cylinder	7. End cap
2. Retaining ring	8. Wear ring
3. Seal ring	9. Piston wear ring
4. "O" ring	10. Seal ring
5. Dust seal	11. "O" ring
6. "O" ring	12. Piston and rod assembly

cap (7) and cylinder (1). Cover hole (H) with sealing compound after the retaining ring is installed to prevent the entrance of water that would rust the end cap and the retaining ring. If removed, install ball-joint (3—Fig. 8) after applying "Loctite 270" to threads. Tighten ball-joint to 80-90 N·m (59-66 ft.-lbs.) torque and ball-joint clamp bolt to 120-160 N·m (88-96 ft.-lbs.) torque. Tighten slotted nut retaining ball-joint in block (8) to 80-140 N·m (60-103 ft.-lbs.) torque, then install cotter pin. Side play of block (8) between bracket arms of axle should be 0-0.2 mm (0-0.008 inch) and is controlled by the thickness of shims (23). Tighten bolts attaching track rods (6) to piston rod to 29-37 N·m (21-27 ft.-lbs.) torque. Tighten slotted nuts attaching rod ends (2) to the steering arms (1) to 50-70 N·m (37-52 ft.-lbs.) torque, then install cotter pins. Adjust position of steering hose elbows for clearance between front support, without hanging down below axle. Make sure hoses do not restrict tipping movement of axle after installation. Refer to paragraph 4 when reinstalling axle.

Four-Wheel-Drive Models

32. R&R AND OVERHAUL. The cylinder is attached between the axle center housing and the steering arm as shown in Fig. 46. To remove the cylinder, disconnect hydraulic hoses at steering cylinder and cover openings to prevent entrance of dirt, then detach both ends of the steering cylinder.

Clamp the cylinder (1—Fig. 47) in a holding fixture and clean sealing compound from elongated hole (H). Rotate the end cap (4) until the end of the retaining ring (2) is located in the hole. Use a small screwdriver or similar instrument to pull and guide the end of the retaining ring out through the elongated hole (H). Turn the end cap (4) and pull the retaining ring from between the end cap and cylinder. One end of the

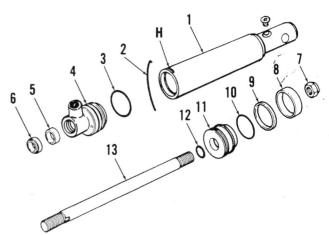

Fig. 47—Exploded view of steering cylinder used on four-wheel-drive tractors.

1. Cylinder	
2. Retaining ring	8. Wear ring
3. "O" ring	9. Seal ring
4. End cap	10. "O" ring
5. Seal	11. Piston
6. Seal	12. "O" ring
7. Nut	13. Piston rod

retaining ring is bent so that end enters a hole in the end cap. After retaining ring (2) is removed, withdraw piston rod (13), piston (11), cap (4) and seals from cylinder (1). Rod end and/or nut (7) can be removed, but if rod end is removed, measure position of old rod end so that new rod end can be installed at same location on rod (13). Remove all old seals and install new seals before reassembling.

Lubricate and assemble steering cylinder by reversing disassembly procedure. Turn end cap (4) in cylinder until the hole in end cap for end of the retaining ring (2) is visible in hole (H). Insert the bent end of retaining ring in hole and turn end cap, drawing retaining ring (2) into groove formed between end cap (4) and cylinder (1). Cover hole (H) with sealing compound after the retaining ring is installed to prevent the entrance of water that would rust the end cap and the retaining ring.

Measure distance between the center of the pivot pin to the center of the ball end when cylinder is completely retracted. Correct distance should be as follows.

Models	Length	
362	418 mm	(16.469 inch)
365, 375	436 mm	(17.178 inch)
390, 390T	436 mm	(17.178 inch)
398	462 mm	(18.203 inch)

Tighten ball-joint to 98-108 N·m (72-80 ft.-lbs.) torque for 362, 365, 375, 390 and 390T models; tighten to 177-196 N·m (130-145 ft.-lbs.) torque for 398 models. Adjust position of steering hose elbows for clearance between front support, without hanging

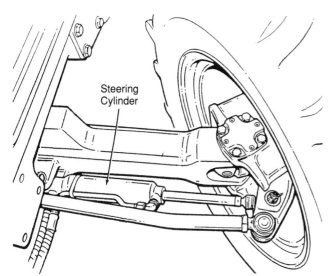

Fig. 46—Installed view of steering cylinder on typical four-wheel-drive tractor.

Steering Cylinder

down below axle. Make sure hoses do not restrict tipping movement of axle after installation.

STEERING CONTROL VALVE

All Models

33. R&R AND OVERHAUL. The steering control valve is located at the base of the steering wheel shaft

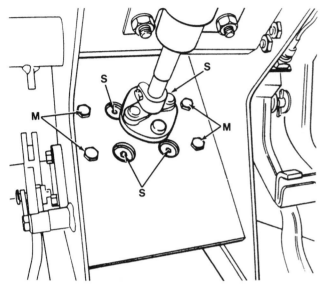

Fig. 48—View of four screws attaching steering control valve. Some models have standard head screws, while other models are fitted with socket head screws as shown.

and removal from tractors with cab requires tractor to be split between the engine and the gearbox as outlined in paragraph 93. Removal is possible without splitting on tractors without cab, but hoses must be disconnected and clearance is very tight. Remove cover from base of steering column, remove the four screws (S—Fig. 48), lower the steering control valve off steering shaft splines, then remove unit from tractor.

On all models, thoroughly clean exterior of unit. Remove the seven cover retaining screws (10—Fig. 49), then remove the cover (9), stator (7), rotor (6), spacer ring (2) and "O" rings (5 and 8). Remove distributor plate (4), drive shaft (1) and "O" ring (3). Hold steering valve vertically and turn valve spool and sleeve to align cross pin (20) parallel to flat side of housing. With cross pin in this position and housing in horizontal position, remove sleeve (22), spool (21), thrust bearing (18) and bearing races (17 and 19) from housing. Remove cross pin (20) from rotary valve and separate spool (21) from sleeve (22). Remove leaf springs (23) from spool.

Do not attempt to remove shock valves from steering cylinder ports (F and G—Fig. 55). The non-return valve in the inlet port (D) should not require servicing either.

If necessary, the pressure relief valve (32 through 36—Fig. 49) can be removed after unscrewing plugs

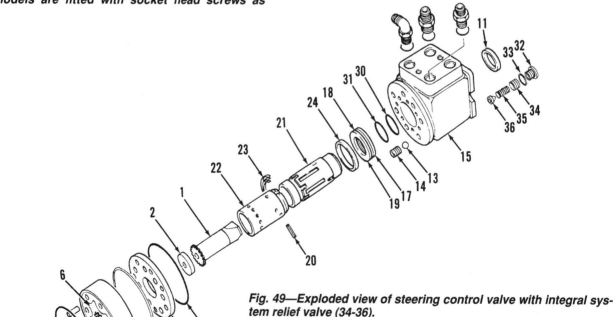

Fig. 49—Exploded view of steering control valve with integral system relief valve (34-36).

1. Drive shaft	11. Oil seal	
2. Spacer	13. Check valve ball	23. Leaf springs (6)
3. "O" ring	14. Check valve plug	24. Retainer
4. Distributor plate	15. Housing	30. Back-up ring
5. "O" ring	17. Bearing race	31. "O" ring
6. Rotor	18. Thrust bearing	32. Plug
7. Stator	19. Bearing race	33. "O" ring
8. "O" ring	20. Cross pin	34. Adjusting plug
9. End cover	21. Valve spool	35. Relief valve spring
10. Cap screw (7)	22. Valve sleeve	36. Relief valve

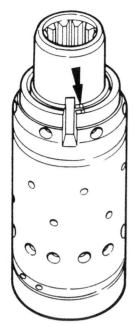

Fig. 50—Marks on spool and sleeve indicated by arrow and slot for centering springs should be aligned as shown when assembling.

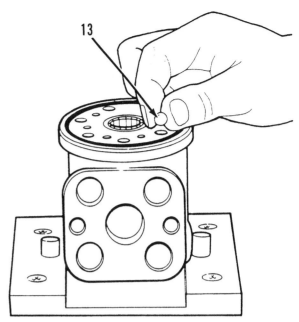

Fig. 52—Check valve ball (13—Fig. 49) should be in location indicated.

(32 and 34); however, pressure must be checked and adjusted if plug (34) is turned. Steering relief pressure is checked as outlined in paragraph 24 or paragraph 25 and adjusted by turning plug (34).

Clean and inspect all parts for excessive wear or other damage and renew parts as necessary. Housing (15), spool (21) and sleeve (22) are not available separately. Use all new "O" rings and seals when reassembling. Lubricate all interior parts with clean steering fluid.

Insert spool (21) into sleeve (22) aligning the two marks (Arrow—Fig. 50) and leaf spring slots. Install leaf springs, with two flat springs (23S) on the outside and two sets of two arched springs (23) in the middle as shown in Fig. 51. Special tool MS63 can be used when installing leaf springs. Insert cross pin (20—

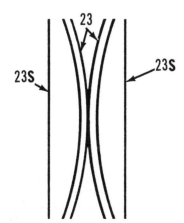

Fig. 51—Arch of the four center leaf springs (23) should be together and the flat springs (23S) should be on the outside as shown.

Fig. 49) into sleeve and spool and install retainer (24) over the leaf springs. Press seal (11) in housing (15) until flush. Install thrust bearing (17, 18 and 19) on spool, making sure that chamfered side of race (17) is away from needle bearing (18). Use special tool MS62A to install the "O" ring and back-up ring into position in housing (15), then carefully insert the assembled and lubricated spool in housing. The cross pin should be kept parallel with the flat surface of housing near the hose connections.

Insert the check valve ball (13) and plug (14) into the hole indicated in Fig. 52. Lubricate the "O" ring (3—Fig. 49) and install in groove of housing, followed by valve plate (4). Make sure that all of the holes in valve plate and housing are aligned. Turn the valve spool assembly so that pin (20) is parallel with the flat port face of housing (15) where hoses attach. Mark a reference line across splined end of rotor shaft (1) parallel with the groove at other end for pin. Be sure that valve spool pin is still aligned with the port face of the housing and install the rotor shaft as shown in Fig. 53. When correctly assembled, one lobe of the rotor will be positioned straight away from the port face (up in Fig. 53) when pin center line is located as shown. Install "O" ring (5—Fig. 49), pump body (7) and pump rotor (6). Install spacer (2), "O" ring (8) and end cover (9). Install the seven retaining screws (10). The one screw with the pin attached should be in location identified as "7" in Fig. 54. Tighten the screws in the order shown in Fig. 54, first to 10-15 N•m (7-10 ft.-lbs.), then to 25-35 N•m (19-26 ft.-lbs.) torque. Install relief valve assembly (32 through 36—Fig. 49) if removed. If setting of plug (34) has been disturbed, refer to paragraph 24 or paragraph 25 for checking and adjusting relief valve pressure. Tighten plugs (24 and 32) to 50 N·m (37 ft.-lbs.) torque.

Installation is reverse of removal procedure. Groove (B—Fig. 55) in splined end of flexible coupling

should be level with top of steering unit drive shaft. If screws (M—Fig. 48) or steering wheel retaining nut (C—Fig. 55) were removed, tighten to 34-37 N·m (25-35 ft.-lbs.) torque. Refer to Fig. 55 for proper attachment of steering hoses (D, E, F and G).

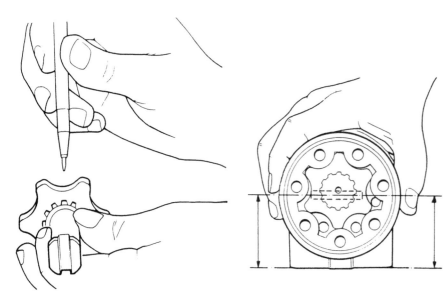

Fig. 53—The center line of pin (20—Fig. 49) should be aligned parallel with the port face of housing as shown when assembling. Center line of pin is below center line of shaft.

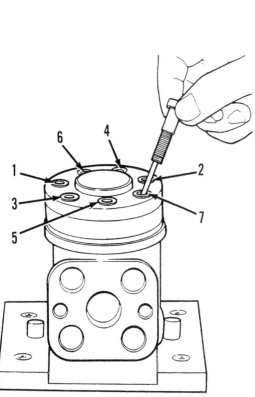

Fig. 54—The one screw with pin attached should be installed in position (7). Numbers indicate the order which screws should be tightened.

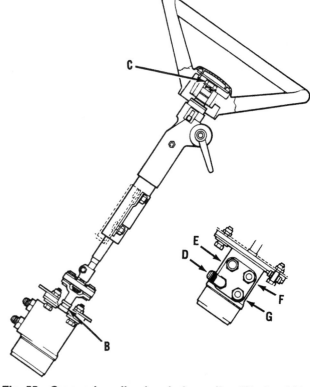

Fig. 55—Groove in splined end of coupling (B) should be level with top of steering shaft. Hose from steering pump should be attached to right angle fitting (D). Hose back to pump or to the oil cooler should be attached to fitting (E). Hose to left steering cylinder should be attached to fitting (F) and hose to right steering cylinder should be attached to fitting (G).

ENGINE AND COMPONENTS

Four-cylinder engines are installed in all models. Refer to Condensed Service Data for original application. The engine serial number consists of up to 13 letters and numbers located as shown in Fig. 56. The numbers and letters constituting the engine serial number can be used to identify the engine family, country of manufacture and the year of manufacture. Refer to Fig. 57 for identifying code.

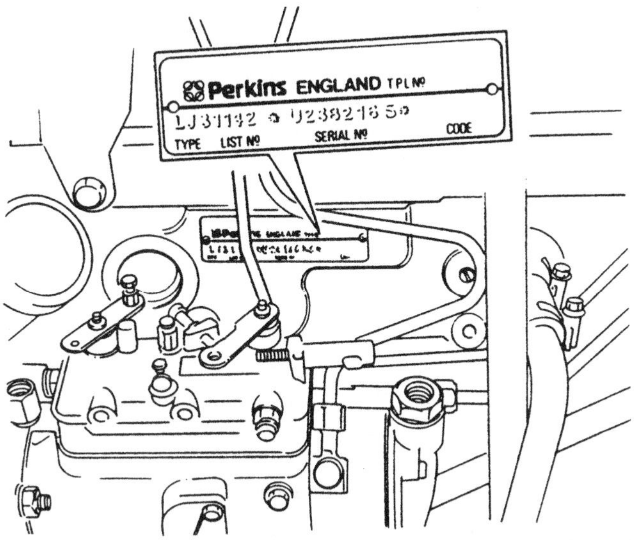

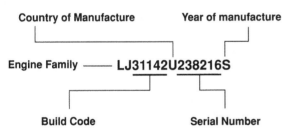

Fig. 56—View of four-cylinder engine showing location of the engine serial number.

Country of Manufacture   Year of manufacture

Engine Family ———— LJ31142U238216S

Build Code   Serial Number

Fig. 57—The first two letters of the engine serial number identifies engine family, the second group is the build code. The letter approximately in the center identifies the country of manufacture and is followed by the engine serial number. The last letter indicates the year of manufacture. The illustration shown above identifies a T4.236 engine (turbocharged, four-cylinder engine with 236 cubic inch displacement), built in the United Kingdom in 1988. The build code (31142) and the serial number is 238216.

Engine family –
LD = 4.236 (4 cyl., 236 cid)
LF = 4.248 (4 cyl., 248 cid)
LJ = T4.236 (Turbocharged,
    4 cyl., 236 cid)
Country of manufacture –
A = Argentina
B = Brazil
C = Australia
D = Germany
E = Spain
F = France
G = Greece
J = Japan
K = Korea
L = Italy
M = Mexico

N = U.S.A.
P = Poland
S = India
T = Turkey
U = United Kingdom
W = Iran
X = Peru
Y = Yugoslavia
Year of manufacture –
N = 1986
P = 1987
S = 1988
T = 1989
U = 1990
V = 1991
W = 1992

FRONT END/ENGINE SPLIT

All Models

34. Block both rear wheels to prevent rolling, remove any front mounted equipment and any front weights. Remove hood, grille and side panels. Drain coolant from radiator and engine block. Remove the air cleaner and exhaust. Disconnect battery ground and front wiring harness to the headlights, horn and fuel tank. If necessary, drain fuel tank, then disconnect fuel feed and return lines. Disconnect top and bottom radiator hoses, hydraulic pipes to and from the oil cooler and hoses to the steering cylinders. On four-wheel-drive tractors, refer to paragraph 9 and remove drive shaft and shield. On all models, wedge between the front support and axle to prevent tipping and support the front axle, radiator and nose from above using a movable hoist. Support rear of tractor under transmission housing, then carefully remove stud nuts and screws attaching front support to the engine. Carefully roll the front axle, radiator and front assembly forward, away from engine as shown in Fig. 58.

Reassemble by reversing the splitting procedure. Refer to paragraph 5 for attaching front support to engine and selecting proper shims. Refer to paragraph 9 for installing drive shaft of four-wheel-drive models.

R&R ENGINE

All Models

35. To remove the engine, first refer to paragraph 34 and split tractor between front support and front of engine. Disconnect wires from engine components and position wires out of the way. Disconnect throttle and stop cables from the fuel injection pump, detach cables from support brackets and move the cable out of the way. Disconnect hydraulic lines between rear of tractor and front that would interfere with the removal of engine and cover all openings. On models with cab, disconnect heater and air conditioning lines if so equipped. Remove gear lever trim plate and boot from models with cab. On all models, attach hoist to the engine, then remove the screws and nuts attaching the engine to transmission housing. Carefully

Fig. 58—View of two-wheel-drive front axle, radiator and associated parts separated from the front of engine. Use care when supporting both front and rear sections to prevent tipping or other unwanted movement.

move the engine forward away from the transmission.

To reinstall engine, reverse the removal procedure and observe the following. Install a M12 alignment stud approximately 100 mm (4 inches) long in each side of the transmission housing to assist in aligning engine to transmission. Turn flywheel to align clutch plate splines with transmission and pto input shaft splines. Install retaining screws and nuts after engine and transmission housing flanges are completely together. Tighten retaining screws and nuts to 115 N·m (85 ft.-lbs.) torque. Reverse removal procedure and refer to paragraph 5 for attaching front support to engine and selecting proper shims. Check and adjust clutch linkage as outlined in paragraph 92. Refer to paragraph 9 for installing drive shaft of four-wheel-drive models.

CYLINDER HEAD

All Models

36. REMOVE AND REINSTALL. To remove cylinder head, first drain cooling system, remove hood and disconnect battery ground cable. Detach exhaust pipe from manifold or turbocharger and inlet pipe from manifold or turbocharger. Remove turbocharger and crossover pipe from models so equipped. Disconnect breather pipe and top radiator hose from all models. Unbolt and remove inlet and exhaust manifolds from cylinder head. Remove the fuel pressure and return lines, unbolt and remove fuel injector assemblies from cylinder head. Remove rocker arm cover, then unbolt and remove the rocker arm assembly. On some models, the cylinder head can be more easily removed if the coolant outlet housing is first removed. Loosen screws retaining cylinder head in reverse of order shown in appropriate Fig. 59 or Fig.

60, then unbolt and lift cylinder head from all models. **Do not use sharp or hard tools between cylinder block and head, and lay removed head on clean flat surface to prevent damage.**

Refer to specific paragraphs for servicing rocker arms, valves, seats and guides. Clean gasket surface of cylinder head and check for distortion or other damage. Check flatness of cylinder head gasket surface using a straightedge and feeler gauge. Cylinder head should be renewed or resurfaced if surface is warped more than 0.08 mm (0.003 inch) across head or 0.15 mm (0.006 inch) from front to rear. No more than 0.30 mm (0.012 inch) can be removed from head, and nozzle protrusion should be less than 4.44 mm (0.175 inch). Standard depth of cylinder head is 102.82-103.58 mm (4.05-4.08 inches).

Different gaskets and fasteners have been used and are available for service. Refer to the following tightening information for the specific engine model, fastener type and gasket used.

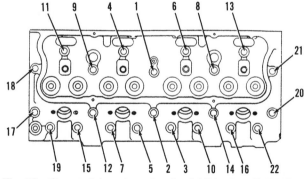

Fig. 59—On models using standard 1/2 inch cylinder head retaining screws, tighten cylinder head retaining screws in the order shown. Removal should be in reverse order shown to prevent damage. Refer to Fig. 60 for correct sequence for late turbocharged models.

Certain cylinder head gaskets should be installed dry, while other gaskets should be coated with sealer. Refer to gasket manufacturer's recommendation that is packed with gasket or, if instructions are not available, install dry unless gasket is copper colored. If cylinder head is retained with only cap screws, studs should be temporarily installed in locations (19 and 22—Fig. 59 or 16 and 21—Fig. 60) to aid alignment of cylinder head and gasket and to prevent sliding while screws are installed. Nuts retaining injection nozzles should be tightened to 18 N·m (14 ft.-lbs.) torque. Coolant outlet (thermostat) housing should be tightened to 13 N·m (10 ft.-lbs.) torque.

On four cylinder models using standard ¹/₂ inch retaining screws, tighten cylinder head retaining screws to 136 N·m (100 ft.-lbs.) torque cold, using the sequence shown in Fig. 59. Retorque retaining screws hot to 120 N·m (88 ft.-lbs.) after running engine until warm and again after operating for 25-50 hours.

Later four cylinder, turbocharged engines use a head gasket marked "TOP FRONT" and "FIT DRY" and cylinder head retaining screws do not require retorquing after operation. Install all cylinder head screws and tighten in sequence shown in Fig. 60 to 120 N·m (88 ft.-lbs.) torque. Recheck torque to make sure that all are tightened evenly, then turn each screw 180° (¹/₂ turn) in the sequence shown in Fig. 60.

On all models, refer to paragraph 41 when installing rocker arm assembly. Check and adjust valve clearance as outlined in paragraph 42. Remainder of reassembly procedure will be reverse of removal procedure.

VALVES AND SEATS

All Models

37. When disassembling the cylinder head, keep valves in order so that they can be reinstalled in

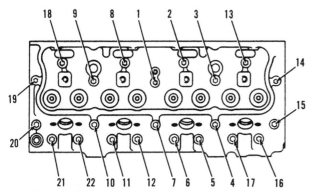

Fig. 60—On late, turbocharged engines, cylinder head retaining screws do not require re-torquing after operation. Tighten cylinder head retaining screws in the order shown. Removal should be in reverse order shown to prevent damage. Refer to Fig. 59 for standard torquing sequence.

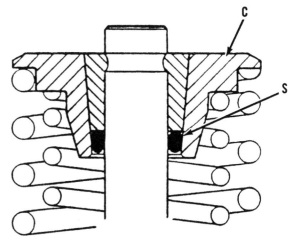

Fig. 61—"O" ring seals (S) are installed as shown on exhaust valves of some models. Valve spring cap (C) is long enough to hold the "O" ring of models so equipped.

original location, if reused. Valve heads are marked with cylinder numbers consecutively from front to rear on some engines. Any replacement valves should be marked before installation.

Inlet and exhaust valves originally seat directly in cylinder head of most models; however, cylinder head may be machined and fitted with replacement seats. Exhaust valve face and seat angles are 45° for all models. Inlet valve face and seat angles are also 45° for all models except 390T and 398 models. Exhaust valves for turbocharged 390T and 398 models also have 45° face and seat angles, but the inlet valves have face and seat angles of 30°. The inlet valve of all models should be recessed 0.89-1.14 mm (0.035-0.045 inch) below surface of cylinder head and should not exceed 1.55 mm (0.061 inch). The exhaust valve of all models should be recessed 1.19-1.45 mm (0.047-0.057 inch) below surface of cylinder head and should not exceed 1.85 mm (0.073 inch). Clearance between valve stem and bore in cylinder head or guide should not exceed 0.13 mm (0.005 inch) for inlet valves; 0.15 mm (0.006 inch) for exhaust valves. Standard valve stem diameter is 9.46-9.48 mm (0.3725-0.3735 inch) for inlet valves of all models; 9.45-9.47 mm (0.3720-0.3728 inch) for exhaust valves of all models. Refer to paragraph 38 for service to valve guides.

Cup type valve stem seals are used on all valves. Late models are also fitted with "O" ring seals (S—Fig. 61) on exhaust valves. The valve spring cap (retainer) is long enough to accommodate "O" ring of models which have "O" ring on exhaust valves.

VALVE GUIDES

All Models

38. Inlet and exhaust valves operate directly in machined bores in the cylinder head of some models.

45

Other models are equipped with valve guides that are pressed into the cylinder head. Clearance between valve stem and bore in cylinder head or guide should not exceed 0.13 mm (0.005 inch) for inlet valves; 0.15 mm (0.006 inch) for exhaust valves. Cup type valve stem seals are installed for both inlet and exhaust valves. Late engines are equipped with "O" ring (S—Fig. 61) on exhaust valves.

Cylinder heads without guides can be resized and valves with oversize valve stems can be installed. Valves with 0.76 mm (0.003 inch), 0.38 mm (0.015 inch) and 0.78 mm (0.030 inch) oversize stems are available for service. Clearance between stem and bore in cylinder head should be 0.02-0.07 mm (0.00075-0.00275 inch) for inlet valves; 0.04-0.08 mm (0.00145-0.00325 inch) for exhaust valves. Be sure to reface valve seats after resizing guide for oversize valve.

New service valve guides can be installed in cylinder heads originally fitted with guides. Do not install valves with oversize stems in cylinder heads fitted with separate valve guides. Stem diameter of new inlet valves is 9.46-9.48 mm (0.3725-0.3732 inch); exhaust valve stem diameter is 9.45-9.47 mm (0.3720-0.3728 inch). Stem to guide clearance should be 0.04-0.09 mm (0.0016-0.0035 inch) for inlet valves; 0.06-0.10 mm (0.0023-0.0039 inch) for exhaust valves. Be sure to reface valve seats after renewing valve guides.

VALVE SPRINGS

All Models

39. Valve springs, retainers and locks are interchangeable for inlet and exhaust valves. Cup type valve stem seals are used on all valves. Late models are also fitted with "O" ring seals (S—Fig. 61) on exhaust valves. The valve spring cap (retainer) is long enough to accommodate "O" ring of models which have "O" ring on exhaust valves. Close wound damper coils should be installed toward cylinder head. Renew springs if discolored, distorted or if they fail to meet the following specifications. Outer spring should exert 17.19-19.01 kg (38-42 lbs.) when compressed to 45.22 mm (1.780 inches). Inner spring should exert 6.55-7.45 kg (14.4-16.4 lbs.) when compressed to 39.7 mm (1.5625 inches).

CAM FOLLOWERS

All Models

40. The mushroom type cam followers operate directly in the machined bores in cylinder block. Cam followers can be removed through bottom of cylinder block after removing oil pan, timing gear cover and

camshaft. Identify cam followers so they can be reinstalled in their original locations if reused.

The 18.99-19.01 mm (0.7475-0.7485 inch) diameter cam followers are available only in standard size and should have 0.04-0.10 mm (0.0015-0.0038 inch) diametral clearance in block bores. Install new cam followers if scored, pitted or excessively worn.

ROCKER ARMS

All Models

41. The rocker arms and shaft assembly can be removed after removing the hood and rocker arm cover. The rocker arms are right and left hand units and should be assembled as shown in Fig. 63.

Rocker shaft diameter is 19.01-19.04 mm (0.7485-0.7495 inch) and the desired diametral clearance between shaft and rocker arms is 0.03-0.09 mm (0.0010-0.0035 inch). Renew shaft and/or rocker arms if clearance exceeds 0.13 mm (0.005 inch) or if either surface of shaft or bushing in rocker arm is rough.

Oil is supplied to the rocker arm shaft through the oil feed connection located in the center of the rocker arm shaft. Two types of oil seal have been used at the oil feed pipe as shown in Fig. 62. If the pipe has two convolutions as shown at left, the "O" ring should be located below the lumps in tube, so that when tube is pushed in place, the "O" ring will roll into place between the convolutions. If the pipe has only one convolution as shown in right side view of Fig. 62, the square section seal ring should be positioned in recess

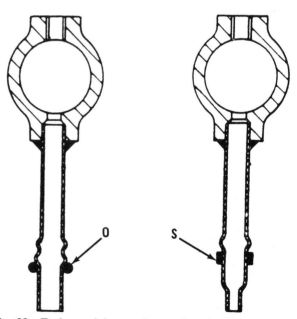

Fig. 62—Early models may be equipped with rocker arm oil feed pipe having two convolutions as shown at left and "O" ring (O) is used. Later models have only one convolution as shown in right side view and a square section seal ring (S) is used.

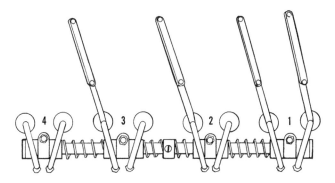

Fig. 63—With TDC timing marks aligned and number 1 (front) piston on compression stroke, adjust the valves indicated. After adjusting these valves, turn crankshaft exactly one revolution (until marks again align) then adjust valves shown in Fig. 64.

of cylinder head, then install rocker shaft and oil delivery tube. Tighten rocker shaft support bracket retaining screws to 41 N•m (30 ft.-lbs.) torque.

VALVE CLEARANCE

All Models

42. Recommended valve clearance is 0.3 mm (0.012 inch) cold for both inlet and exhaust valves. Static setting of all valves can be made from two crankshaft positions using the following procedure:

Remove the timing window plug from the left side of flywheel adapter housing and turn the flywheel until "TDC 1" mark on flywheel is aligned in center of hole. Check valves for front cylinder. Both valves should be closed if front cylinder is on compression stroke. If front two valves are tight (open), turn crankshaft exactly one complete revolution to set front piston on compression stroke.

If not equipped with hole in flywheel adapter or mark on flywheel, turn flywheel in normal direction of rotation until the exhaust valve for the front cylinder is just about closed and the inlet valve is just

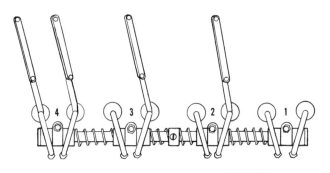

Fig. 64—With TDC timing marks aligned and number 1 (front) piston on exhaust stroke, adjust the valves indicated.

beginning to open. If not equipped with marks, mark crankshaft pulley so that crankshaft can be rotated exactly one complete revolution from this position after adjusting the first set of valves.

With front cylinder set at TDC on compression stroke, adjust the four valves indicated in Fig. 63. Turn the crankshaft exactly one complete revolution and adjust the four valves indicated in Fig. 64.

VALVE TIMING

All Models

43. Timing gears are correctly timed if marks are properly aligned as outlined in paragraph 50. Valve timing is not adjustable and gears can only be improperly timed by improper assembly or damage to the gear teeth. Refer to paragraph 44 for removing the timing gear cover, then to paragraph 50 for checking and setting the timing marks.

TIMING GEAR COVER

All Models

44. To remove the timing gear cover, refer to paragraph 34 and remove the front axle, support and associated parts. Remove accessory drive belts, alternator, fan blades and crankshaft pulley. Remove screws attaching the timing gear cover to the front of engine. Be careful to make sure that all of the screws are removed.

Press (do not pry) old crankshaft oil seal from cover bore, then press new seal into position with spring loaded lip toward inside. Two different types of oil seals have been used. Early models are black (nitrile) and crankshaft is equipped with an oil thrower. Later models are red (silicone) and a spacer is used in place of the oil thrower. **Do not use oil thrower with red (silicone) seal.** Both seals are directional and stamped with an arrow to indicate the direction of rotation of the engine crankshaft. Press new seal into bore until front surface of seal case is 9.65-9.91 mm (0.38-0.39 inch) from front surface of timing gear cover.

To install timing gear cover, first install a new seal into cover, then position gasket on front of engine. Position cover and install retaining screws loosely. Center timing gear cover around the crankshaft using special centering tool (PD162 available from Massey-Ferguson) or equivalent, then tighten the retaining screws. If special centralizing tool is not available, the crankshaft pulley can be used carefully, but damage to the soft lip of the seal is likely. Install the crankshaft pulley, aligning the punch mark on front of pulley hub with the stamped line on front of crankshaft. The crankshaft pulley may be retained by either one screw or three screws. Refer to the following recommended torque values and make sure that screw or screws are tightened correctly.

One Screw, 42.4 mm
   (1.67 inches) long . . . .   406 N·m
                            (300 ft.-lbs.)
One Screw, 38.1 mm
   (1.5 inches) long—
   Cadmium plated
   (silver) . . . .   325 N·m
                            (240 ft.-lbs.)
   Phosphated (black) . . . .   390 N·m
                            (285 ft.-lbs.)
Three Screws . . . .   95 N·m
                            (70 ft.-lbs.)

Complete assembly by reversing the disassembly procedures.

TIMING GEARS

All Models

45. Before attempting to remove any of the timing gears, first remove the rocker arm cover and the rocker arms to avoid possible damage to pistons or valve train. The valves can (and probably will) hit the pistons if either the camshaft or crankshaft is turned independently of the other. The timing marks shown in Fig. 65 align only once every 18 revolutions due to the odd number of teeth on the idler gear.

To remove the timing gears or align marks on gears, refer to paragraph 34 and remove the front axle, support and associated parts, then remove the timing gear cover as outlined in paragraph 44. Backlash should be at least 0.08 mm (0.003 inch) between the idler and any other gear. Refer to paragraphs 46, 47, 48 and 49 for service information regarding each of the four gears. With rocker arms removed, crankshaft gear, camshaft gear and injection pump gear can be rotated as necessary before reinstalling idler gear with all timing marks aligned.

46. IDLER GEAR. Backlash should be at least 0.08 mm (0.003 inch) between the idler and any other gear. Gear should have 0.20-0.41 mm (0.008-0.016 inch) end play on shaft. The 63 tooth idler gear is retained on idler shaft by three screws and a washer. Thrust washer should be 3.02-3.10 mm (0.119-0.122 inch) thick and idler shaft diameter should be 50.70-50.72 mm (1.996-1.997 inches). Gear to shaft diametral clearance should be 0.07-0.12 mm (0.0028-0.0047 inch). The two flanged bushings in idler gear must be reamed after installation to inside diameter of 50.79-50.82 mm (1.9998-2.0007 inches).

Make sure that timing marks on all gears are aligned as shown in Fig. 65 when idler gear is installed. Tighten retaining screws to 41 N·m (30 ft.-

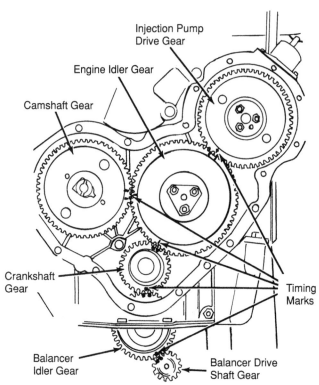

Fig. 65—View of timing marks aligned on four-cylinder engine.

lbs.) torque, then bend tabs of lock plate against a flat of the screws to prevent loosening.

47. CAMSHAFT GEAR. Remove the rocker arms and idler gear before removing the camshaft gear. The camshaft gear (Fig. 65) is pressed and keyed to the shaft and retained by a special cap screw, tab washer and retaining plate. Remove the special retaining screw, then use a suitable puller to pull gear from the shaft using the two tapped holes in gear. **Gears stamped "M" have metric thread in puller holes.**

Camshaft end play should be 0.10-0.41 mm (0.004-0.016 inch) and is controlled by thrust washer located behind the timing gear housing.

When reinstalling camshaft gear, use the retaining plate and special cap screw to draw gear into position on shaft. **Do not attempt to drive gear onto shaft, because plug in rear of cylinder block will be damaged.** Tighten special gear retaining cap screw to 68 N·m (50 ft.-lbs.) torque, then lock in place by bending tab of washer against flat of screw head. Refer to paragraph 50 for timing the gears.

48. CRANKSHAFT GEAR. The crankshaft gear is keyed in place and fits the shaft with a transitional fit, 0.03 mm (0.001 inch) tight to 0.03 mm (0.001 inch) loose. If the old gear is loose on shaft, it may be possible to pry gear from shaft, but a puller is usually required. If necessary to remove crankshaft gear with

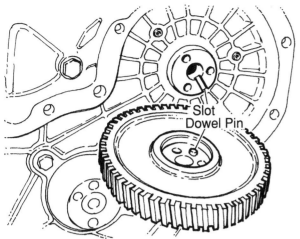

Fig. 66—Correct installation of injection pump drive gear is simplified by the dowel pin which fits in machined slot in pump drive shaft.

a puller, the oil pan, timing gear housing and balancer assembly must first be removed.

When installing gear, be sure that marked tooth is toward front. Install washers and oil slingers as originally fitted behind and in front of crankshaft gear.

49. INJECTION PUMP DRIVE GEAR. The injection pump drive gear is retained to the pump adapter by either three cap screws or a center retaining nut. When installing the gear, align dowel pin (Fig. 66) with slot in adapter hub or drive key with slot, then install the retaining screws or nut. The injection pump drive gear and adapter are supported by the injection pump rotor bearings.

50. TIMING THE GEARS. The crankshaft gear has 28 teeth, the camshaft gear has 56 teeth and the injection pump drive gear has 56 teeth, but because the idler gear has 63 teeth, marks on the four gears will align only every 18 crankshaft revolutions. To align the marks, it is easier and faster to turn the crankshaft until the marked teeth of the camshaft and injection pump gears both mesh with the idler gear. The marked tooth of the crankshaft gear will also be in mesh with the idler gear at the same time. Remove the idler gear, then reinstall with all of the marks aligned as shown in Fig. 65.

Do not turn the crankshaft or camshaft with the idler gear removed, because the valves can (and probably will) hit the pistons if either the camshaft or crankshaft is turned independently of the other. If it is necessary to turn either the camshaft or crankshaft to align timing marks, first remove the rocker arm cover and the rocker arms to avoid possible damage to pistons or valve train.

If rocker arms have been removed, turn the crankshaft, camshaft and injection pump gears to approxi-

mately the positions shown in Fig. 65, then install the idler gear with all of the marks aligned.

The balance shaft gear must be timed to the crankshaft gear as shown in Fig. 65, but timing is possible as described in paragraph 59 without removing the timing gear cover.

TIMING GEAR HOUSING

All Models

51. To remove the timing gear housing, first remove the timing gear cover as outlined in paragraph 44, the rocker arm cover and the rocker arms assembly. Remove the idler gear as outlined in paragraph 46, camshaft gear as outlined in paragraph 47 and the injection pump drive gear as outlined in paragraph 49. Remove the injection pump and the steering pump if not already removed. Remove the front four oil pan screws, if the oil pan is not removed. Remove the screws securing the timing gear housing to front face of cylinder block, then lift housing away as shown in Fig. 67.

The timing gear housing must be removed before camshaft can be withdrawn. Install by reversing removal procedure. Oil pan must be installed and tightened before tightening the screws retaining the timing gear housing to front face of engine block.

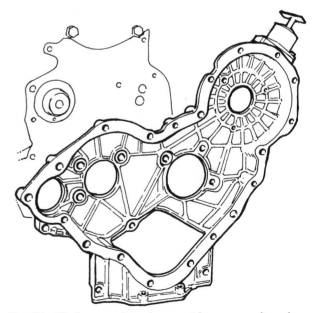

Fig. 67—Timing gear housing must be removed as shown before camshaft thrust washer and camshaft can be removed.

CAMSHAFT

All Models

52. To remove the camshaft, first remove the timing gear housing as outlined in paragraph 51 and the oil pan. Secure cam followers in their uppermost position, remove the fuel lift pump, then withdraw camshaft and front thrust washer. Refer to paragraph 40 for service to cam followers which can be removed after camshaft.

Camshaft end play should be 0.10-0.41 mm (0.004-0.016 inch) and is controlled by thrust washer located behind the timing gear housing. Install new thrust washer if camshaft end play exceeds 0.51 mm (0.020 inch). The thrust washer is located by a dowel pin in block (Fig. 68). The thrust washer should be 5.47-5.54 mm (0.216-0.218 inch) thick. Depth of recess for the thrust washer and stand out differ between models. Block recess for earliest models is 3.86-3.91 mm (0.152-0.154 inch) and thrust washer should protrude 1.53-1.68 mm (0.062-0.066 inch) beyond front face of cylinder block. The block recess for later models is 4.75-4.83 mm (0.187-0.190 inch) and thrust washer should protrude 0.66-0.79 mm (0.026-0.031 inch) beyond front face of cylinder block. Recess in block for the latest models is 5.46-5.53 mm (0.215-0.218 inch) and thrust washer should be nearly flush with front face of cylinder block. Actual position of the thrust washer for latest models should be 0.05 mm (0.002 inch) below front face of cylinder block to 0.07 mm (0.003 inch) above front face of cylinder block.

Specified cam lift is 7.62-7.70 mm (0.300-0.303 inch). The camshaft operates in three bearings. The front bearing is fitted with a presized renewable bushing (Fig. 68), while the remaining camshaft journals operate directly in machined bores in engine block. Refer to the following for journal specifications.

Front Journal

Journal OD	50.71-50.74 mm (1.9965-1.9975 inches)
Bearing ID	50.80-50.83 mm (2.000-2.001 inches)
Clearance	0.06-0.11 mm (0.0025-0.0045 inch)

Center Journal

Journal OD	50.46-50.48 mm (1.9865-1.9875 inches)
Bearing ID	50.55-50.60 mm (1.990-1.992 inches)
Clearance	0.06-0.14 mm (0.0025-0.0053 inch)

Rear Journal

Journal OD	49.95-49.98 mm (1.9665-1.9675 inches)
Bearing ID	50.04-50.09 mm (1.970-1.972 inches)
Clearance	0.06-0.14 mm (0.0025-0.0053 inch)

ROD AND PISTON UNITS

All Models

53. Connecting rod and piston units are removed from above after removing cylinder head as outlined in paragraph 36, oil pan and engine balancer assembly. Remove carbon and wear ridge from top of cylinder bore, unbolt and remove bearing cap, then push rod and piston assembly out of cylinder. Make sure that cylinder number (C—Fig. 69) is stamped on both rod and cap on side away from camshaft. The side of rod and cap that is marked with cylinder number is also the side which has tangs of bearing inserts.

Notice that two different types of connecting rod nuts have been used and each has different tightening torques. Early models were originally fitted with cadmium plated (bright finish) nuts which should be tightened to only 102 N·m (75 ft.-lbs.) torque. Later nuts are phosphated, have a dull black finish and should be tightened to 129 N·m (95 ft.-lbs.) torque.

It is important that the top of piston is correct distance from the top face of cylinder block when assembled. Pistons are available with top surfaces different distances from piston pin and also "untopped," which require machining the piston top to the correct height. Refer to paragraph 54.

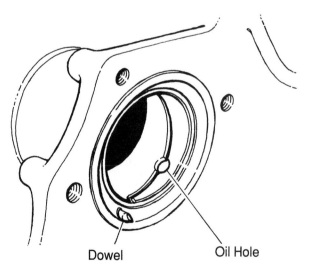

Fig. 68—Front face of engine block with camshaft removed, showing front camshaft bushing and thrust washer locating dowel pin. Other camshaft bores do not have renewable bushings.

Dowel   Oil Hole

Fig. 69—View of typical piston and connecting rod correctly assembled with "F" or "FRONT" mark (A), cylinder number (B) and rod correlation numbers (C).

PISTONS, RINGS AND SLEEVES

All Models

54. The aluminum alloy, cam ground pistons are supplied in standard size only and are available in a kit consisting of piston, pin and rings for one cylinder. The combustion chamber is offset in piston toward injection nozzle side of engine. Piston is marked "FRONT" or "F" as shown at (A—Fig. 69) to facilitate assembly. Additionally, the offset combustion chamber should be toward same side as numbers stamped on connecting rod and cap, which should be installed in engine away from camshaft side.

It is important that the top of piston is a specific distance from the top face of cylinder block when assembled. On models 362, 365 and 375, the piston should be within the range of 0.61 mm (0.024 inch) below to 0.41 mm (0.016 inch) above the surface of the cylinder block. On naturally aspirated models 383 and 390, the piston should be within the range of 0.25 mm (0.0098 inch) below to 0.08 mm (0.0.0031 inch) above the surface of the cylinder block. On turbocharged models 390T and 398, the piston should be within the range of 0.61 mm (0.024 inch) below to 0.41

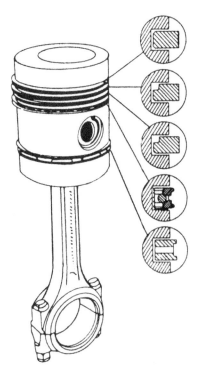

Fig. 70—Drawing of five-ring piston with cross section of piston rings typical of use in naturally aspirated A4.236 engines for 362, 365 and 375 models.

mm (0.016 inch) above the surface of the cylinder block.

Pistons are available with top surfaces different distances from piston pin and also "un-topped," which require machining the piston top to the correct height. Install the connecting rod and piston assembly to the crankshaft and in cylinder block, but without piston rings. Mark the top edge of cylinder block with piston at TDC, then remove and machine top of piston as required. After machining top of piston, top inner edge of the combustion chamber in piston should be rounded slightly and top of piston should be stamped with marks (A and B—Fig. 69) indicating cylinder number and front.

Five piston rings (Fig. 70) are used in naturally aspirated engines used in 362, 365 and 375 models. The first (top) ring is chrome plated and if marked, should be installed with "T" or "TOP" side up. The second and third rings are alike and are internally stepped. The internal groove should be installed toward top of piston. The fourth, oil scraper ring is non-directional with a coil spring expander. Install the expander in the groove first and join ends with the locating wire, then install ring over the expander, making certain that joint of expander is 180° from (opposite) gap of ring. The bottom (fifth), oil control ring is non-directional and may be installed with either side up. Piston and ring specifications are as follows.

Five Ring Pistons

Ring Width –

Top . . . . 2.36-2.38 mm
(0.0928-0.0938 inch)

Second and Third . . . . 2.36-2.38 mm
(0.0928-0.0937 inch)

Fourth & Bottom. . . . . 6.33-6.35 mm
(0.249-0.250 inch)

Ring Side Clearance in Groove –

Top . . . . 0.05-0.10 mm
(0.0019-0.0039 inch)

Second and Third . . . . 0.05-0.10 mm
(0.0019-0.0039 inch)

Fourth & Bottom. . . . . 0.06-0.11 mm
(0.0025-0.0045 inch)

Ring End Gap –

Top . . . . 0.41-0.86 mm
(0.016-0.034 inch)

Second and Third . . . . 0.30-0.76 mm
(0.012-0.030 inch)

Fourth . . . . 0.30-0.43 mm
(0.012-0.017 inch)

Bottom. . . . . 0.30-0.76 mm
(0.012-0.030 inch)

Three piston rings (Fig. 71) are used in naturally aspirated engines used in 383 and 390 models. The molybdenum first (top) ring is barrel faced, internally stepped and externally chamfered. The second ring is taper faced, internally stepped and externally stepped. The internal groove (step) should be installed toward top of piston. The bottom (third), oil control ring is non-directional with a coil spring expander. Install the expander in lower groove first and join ends with the locating wire. Install the ring over the expander, making certain that joint of expander is 180° from (opposite) gap of ring. Piston and ring specifications are as follows.

Three Ring Pistons (Naturally Aspirated)

Ring Width –

Top . . . . 2.46-2.49 mm
(0.097-0.098 inch)

Second . . . . 2.46-2.49 mm
(0.097-0.098 inch)

Bottom. . . . . 4.96-4.99 mm
(0.1954-0.1964 inch)

Ring Side Clearance in Groove –

Top . . . . 0.04-0.07 mm
(0.0017-0.0027 inch)

Second . . . . 0.04-0.07 mm
(0.0017-0.0027 inch)

Bottom. . . . . 0.05-0.10 mm
(0.002-0.0038 inch)

Ring End Gap –

Top . . . . 0.41-0.86 mm
(0.016-0.034 inch)

Second . . . . 0.41-0.86 mm
(0.016-0.034 inch)

Bottom. . . . . 0.41-0.86 mm
(0.016-0.034 inch)

Three piston rings (Fig. 72) are used in turbocharged 390T and 398 models. The first (top) ring is tapered and the face is chrome plated. The second ring has a tapered face and ring must be installed with "T" or "TOP" mark toward top of piston. The bottom (third), oil control ring is non-directional with a coil spring expander. Install the expander in lower groove first and join ends with the locating wire. Install the ring over the expander, making certain that joint of expander is 180° from (opposite) gap of ring. Piston and ring specifications are as follows.

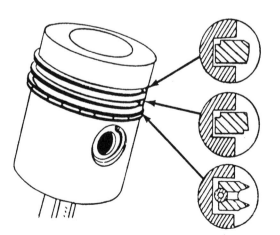

Fig. 71—Drawing of three-ring piston with cross section of piston rings typical of use in naturally aspirated A4.248S engines for 383 and 390 models.

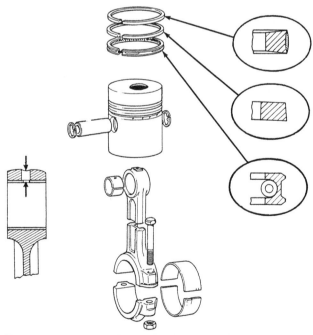

Fig. 72—Drawing of three-ring piston with cross section of piston rings typical of use in turbocharged AT4.236 engines for 390T and 398 models.

Three Ring Pistons

Ring Width –

Top . . . . 2.36-2.37 mm
(0.0928-0.0935 inch)

Second . . . . 2.35-2.38 mm
(0.0927-0.0937 inch)

Bottom . . . . 4.72-4.73 mm
(0.1860-0.1865 inch)

Ring Side Clearance in Groove –

Top . . . . Tapered

Second . . . . 0.05-0.08 mm
(0.0020-0.0033 inch)

Bottom . . . . 0.03-0.08 mm
(0.001-0.0033 inch)

Ring End Gap –

Top . . . . 0.25-0.61 mm
(0.010-0.024 inch)

Second . . . . 0.25-0.66 mm
(0.010-0.027 inch)

Bottom . . . . 0.25-0.79 mm
(0.010-0.031 inch)

Original cylinder bore diameter is 98.48-98.50 mm (3.877-3.878 inches) for A4.236 and AT4.236 engines used in 362, 365, 375, 390T and 398 models. Original cylinder bore diameter is 101.05-101.07 mm (3.9785-3.9795 inches) for A4.248S engines used in 383 and 390 models. Install new liner if taper, out-of-round or wear exceeds 0.2 mm (0.008 inch). Suitable sleeve removal and installing tools are required to renew sleeves. Thoroughly clean and inspect cylinder block bores and new sleeves before installing, because even the slightest burr or dirt can cause distortion of new sleeve as it is pressed into block. Measure height of liner (sleeve) flange and depth of block recess. Cylinder liner flange should be from 0.10 mm (0.004 inch) below surface of block to 0.10 (0.004 inch) above block surface. Shims are available which may be installed under flange if necessary to raise sleeve. Make sure that block recess is clean. Standard depth of block recess is 3.81-3.91 mm (0.150-0.154 inch).

Lubricate cylinder block bores with new engine oil just prior to installing new sleeves. Flanged, prefinished, service sleeves should fit block bores from 0.025 mm (0.001 inch) loose to 0.025 mm (0.001 inch) tight. Correct installed diameter of prefinished service liners for A4.236 and AT4.236 engines is 98.50-98.52 mm (3.878-3.879 inches). Correct installed diameter of prefinished service liners for A4.248S engines is 101.09-101.12 mm (3.980-3.981 inches). Allow sufficient time for cylinder liner to stabilize, then check installed diameter of sleeves carefully at several locations from top to bottom of bore and in several different locations across bore.

Unfinished cylinder liners similar to those used in original production are available for service. Unfinished liners are 0.03-0.08 mm (0.001-0.003 inch) interference fit in cylinder block bores and require heavy pressing equipment. Accurate boring equipment should be used to machine installed sleeve to original cylinder bore diameter of 98.48-98.50 mm (3.877-3.878 inches) for A4.236 and AT4.236 engines; 101.05-101.07 mm (3.9785-3.9795 inches) for A4.248S engines. Sleeves with oversize outside diameter are also available for service. Standard block bore is 103.19-103.22 mm (4.0625-4.0635 inches) for A4.236 and AT4.236 engines. Standard block bore is 104.19-103.22 mm (4.0625-4.0635 inches) for A4.236 and AT4.236 engines.

PISTON PINS

All Models

55. The full floating piston pin is retained in piston bosses by snap rings.

On naturally aspirated A4.236 and A4.248S engines, pin diameter is 34.92-34.930 mm (1.3748-1.3750 inches). The piston pin is a transitional fit in piston and should have 0.019-0.043 mm (0.00075-0.00170 inch) clearance in bushing at top of connecting rod. To remove piston pin, remove snap rings from both ends, then heat the piston to 40-50° C (100-120° F). Pin should slide out after heating. Do not attempt to press pin in or out before heating piston.

Pin diameter is 38.095-38.100 mm (1.4998-1.5000 inches) diameter for turbocharged A4.236 engines. The piston pin should have 0-0.010 mm (0-0.0004 inch clearance in piston and should have 0.019-0.043 mm (0.00075-0.00170 inch) clearance in bushing at top of connecting rod. To remove piston pin, remove snap rings from both ends, then heat the piston to 40-50° C (100-120° F). Pin should slide out after heating. Do not attempt to press pin in or out before heating piston.

Bushing in top of rod is renewable. Be sure that oil hole in bushing is aligned with oil hole in top of rod bore. Bushing must be reamed to provide correct clearance for pin. Refer to paragraph 56 for installation of new pin bushing.

CONNECTING RODS AND BEARINGS

All Models

56. Connecting rod crankpin bearings are steel backed, aluminum/tin or bronze/lead faced type renewable from below after removing the oil pan, engine balancer and rod bearing caps. Projections (tangs) on bearing insert should engage milled slots in rod and cap. The tangs (T—Fig. 73) on bearing inserts must both be on same side of engine and away from camshaft. If new connecting rod is installed, cylinder number (C) should be stamped on rod and cap.

Bushing in top of rod for piston pin is renewable. Be sure that oil hole in bushing is aligned with oil

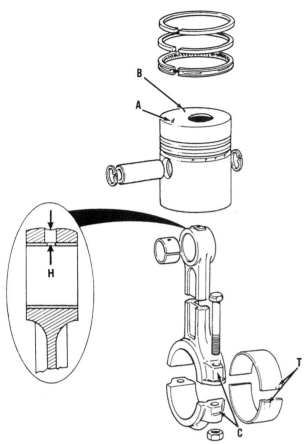

Fig. 73—Drawing of connecting rod showing relative position of cylinder numbers on rod and cap (C), tangs (T) on bearing inserts, front mark (A) on piston and cylinder number (B) on piston. Hole (H) in piston pin bushing must align with hole in connecting rod.

CRANKSHAFT AND BEARINGS

All Models

57. The crankshaft is supported in five steel backed, aluminum/tin or lead/bronze faced type, renewable bearing inserts. Projections (tangs) on bearing insert must engage milled slots in block and main bearing cap. Tangs of inserts in block and cap must both be on same side of engine.

To remove the crankshaft, first drain coolant and engine oil, then remove engine as outlined in paragraph 35. Remove oil pan, timing gear cover, timing gear housing and engine balancer. Remove clutch, flywheel, engine adapter plate and crankshaft rear oil seal. Remove connecting rod caps and main bearing caps, then lift crankshaft from cylinder block.

The upper and lower halves of main bearing inserts are not interchangeable. The upper insert halves are slotted to provide pressure lubrication to crankshaft and connecting rods. Inserts are interchangeable in pairs (upper and lower halves) for all journals except the center main bearing. Main bearing inserts are available in standard size and 0.25, 0.50 and 0.75 mm (0.010, 0.020 and 0.030 inch) undersizes. If crankshaft main journal is resized, journal fillet radii should be 3.68-3.96 mm (0.145-0.156 inch).

Crankshaft end play is controlled by renewable thrust washers at front and rear of center main bearing. The cap half is prevented from turning by the tab which fits in a machined notch of cap. Block half of washer can be rolled from position when cap is removed. Recommended crankshaft end play is 0.05-0.38 mm (0.002-0.015 inch). Thrust washers are available in 0.18 mm (0.007 inch) oversize as well as standard thickness. One set (top and bottom) of oversize thrust washers may be installed on one side (front or rear) in combination with a standard size set on other side to provide oversize adjustment of 0.18 mm (0.007 inch). Two sets of oversize thrust washers may be used to provide adjustment of 0.36 mm (0.014 inch).

Check crankshaft against the values which follow.

hole in top of rod bore while pressing new bushing into position. Size new, installed bushing to 34.94-34.96 mm (1.37575-1.37650 inch) for naturally aspirated A4.236 and A4.248S engines. Size new, installed bushing to 38.12-38.14 mm (1.50075-1.50150 inch) for turbocharged AT4.236 engines. Clearance between piston pin and correctly sized bushing should be 0.019-0.043 mm (0.00075-0.00170 inch). Pin bushing must also be parallel to bore for crankshaft crankpin. Connecting rod length between centers is 219.05-219.10 mm (8.624-8.626 inches) for all of these four cylinder engines.

On all models, crankshaft crankpin diameter is 63.47-63.49 mm (2.4988-2.4996 inches) new. Bearing inserts are available in standard size and 0.25, 0.50 and 0.75 mm (0.010, 0.020 and 0.030 inch) undersizes. If crankpin is resized, crankpin fillet radii should be 3.68-3.96 mm (0.145-0.156 inch). Rod bearing diametral clearance on crankpin should be 0.03-0.08 mm (0.0012-0.0031 inch) and rod should have 0.21-0.37 mm (0.0085-0.0148 inch) end play on crankpin.

Main Journal Standard
   Diameter. . . . . 76.16-76.18 mm
                           (2.9984-2.9992 inches)
Main Journal Fillet Radii . . . . 3.68-3.96 mm
                           (0.145-0.156 inch)
Main Bearing Clearance . . . . 0.05-0.11 mm
                           (0.0018-0.0042 inch)
Crankshaft End Play. . . . . 0.05-0.38 mm
                           (0.002-0.015 inch)
Crankpin Standard Diameter. . . . . 63.47-63.49 mm
                           (2.4988-2.4996 inches)
Crankpin Fillet Radii . . . . 3.68-3.96 mm
                           (0.145-0.156 inch)

Rod Bearing Clearance on Crankpin . 0.03-0.08 mm
(0.0012-0.0031 inch)
Rod End Play on Crankpin . . . . 0.21-0.37 mm
(0.0085-0.0145 inch)

To reinstall crankshaft, reverse removal procedure. Notice that main bearing caps are numbered 1 through 5 with number 1 cap at front of engine. The caps are also marked with the serial number of the engine. Make certain thrust washers are installed with grooved side toward crankshaft. Tighten main bearing cap screws to 244 N•m (180 ft.-lbs.) torque.

ENGINE BALANCER

All Models

58. The balancer assembly consists of two shafts and weight assemblies which rotate in opposite directions at twice crankshaft speed. If properly timed, the balancer weights will be positioned at their lowest point (flat surfaces of weights toward top) each time the pistons are at TDC and BDC of their stroke.

The balancer unit is driven by the crankshaft gear through an idler gear attached to the balancer frame. The engine oil pump (16—Fig. 74) is installed in the center of balancer frame and is driven by the balancer drive shaft.

59. REMOVE AND REINSTALL. The balancer assembly can be removed after removing the oil pan and balancer mounting screws.

To reinstall and "time" balancer to the crankshaft, first turn engine crankshaft until No. 1 and 4 pistons are both at TDC. Let balance weights hang downward (flat surface toward top) as shown in Fig. 76. Flat surface of balance weights (2 and 3) should be aligned and parallel to the mounting surface of block. Position the balancer unit on bottom of cylinder block, meshing idler gear (9) with crankshaft gear, then tighten balancer mounting screws to 36 N•m (49 ft.-lbs.) torque.

Timing marks may be located on idler gear and balance shaft drive gear as shown in Fig. 65; however, these marks can be used to time the balancer only if timing gear cover has been removed. Balance weights

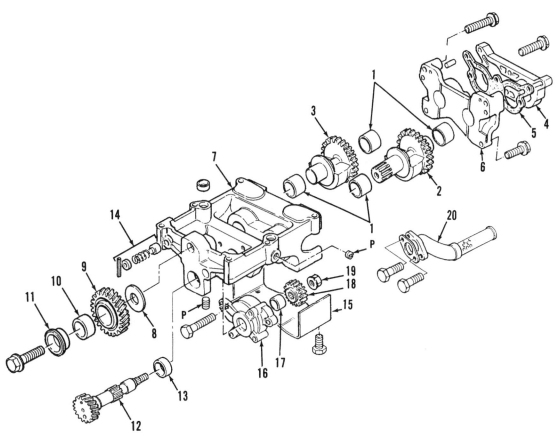

Fig. 74—Exploded view of balancer assembly used on four-cylinder engines. Plugs (P) should not be removed.

1. Bushings	6. End cover	11. Hub	16. Oil pump assy.
2. Balancer weight (driven)	7. Balancer frame	12. Drive shaft	17. Needle bearing
3. Balancer weight (drive)	8. Thrust washer	13. Needle bearing	18. Balancer drive gear
4. Oil transfer plate	9. Idler gear	14. Oil pressure relief valve	19. Locknut
5. Gasket	10. Needle bearing	15. Cover plate	20. Pump inlet

may hit connecting rod causing serious damage if incorrectly assembled to engine.

60. OVERHAUL. Refer to Fig. 74 for an exploded view of balancer assembly. To disassemble the removed balancer, remove the cap screw retaining idler gear (9) and hub (11) to balancer frame. Hold the drive shaft gear (12), then remove nut (19) from end of shaft. Withdraw drive shaft and balancer drive gear (18). Remove cover plate (15). Unbolt and remove oil pump (16) from the frame. Remove cap screws retaining oil transfer plate (4) and end cover (6), then separate plate and cover from frame. Remove the balance weights (2 and 3). Press bushings (1) from end cover and frame if necessary.

Do not remove plugs (P—Fig. 74). These plugs determine the direction of oil flow and must be assembled in correct locations. Additionally, plugs are sealed in position and removal may damage plug or threaded bore. Lubricating oil passages should, however, be cleaned and flushed. Examine all parts for wear or damage and renew as necessary. Refer to the following specifications.

Drive Shaft (12)
Journal OD –
    Front . . . . 28.562-28.575 mm
          (1.124-1.125 inches)
    Rear . . . . 23.787-23.800 mm
          (0.936-0.937 inch)
End Play . . . . 0.10-0.30 mm
          (0.004-0.012 inch)

Balance Weights (2 & 3)
Shaft OD . . . . 38.054-38.069 mm
          (1.498-1.499 inches)
Bushing (1) ID . . . . 38.133-38.174 mm
          (1.501-1.503 inches)
Shaft to Bushing,
    Diametral Clearance . . . . 0.064-0.120 mm
          (0.0025-0.0047 inch)
Gear Backlash . . . . 0.10-0.265 mm
          (0.004-0.010 inch)
End Play . . . . 0.186-0.377 mm
          (0.0073-0.0148 inch)

Idler Gear (9)
Hub OD . . . . 38.090-38.100 mm
          (1.4996-1.500 inches)
Thrust Washer (8) Thickness . . . . 4.14-4.29 mm
          (0.163-0.169 inch)
End Play . . . . 0.08-0.23 mm
          (0.003-0.009 inch)

When renewing drive shaft bearings (13 and 17—Fig. 74), press new front bearing (13) into balancer housing until front of bearing is 2.5-3.0 mm (0.098-0.118 inch) below machined surface of housing. Press

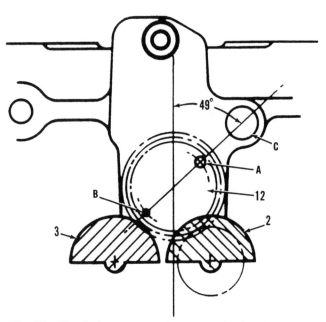

Fig. 75—The balancer must be correctly timed while assembling and installing. Align holes (A and B) in drive gear (12) with hole (C), then install balance weights and shafts (2 and 3) with weights up and flats aligned as shown.

rear bearing (17) into housing until rear of bearing is 2.00-2.30 mm (0.079-0.118 inch) below machined surface of housing.

When renewing balance weight bushings (1), press new bushings into balancer housing until rear ends of bushings are 3.25-3.30 mm (0.128-0.130 inch) below machined surface of housing bores. Press bushings into end cover (6) until front ends of bushings are 3.25-3.30 mm (0.128-0.130 inch) below machined surface of cover. Refer to paragraph 64 for inspection and service to oil pump and to paragraph 65 for inspection of oil pressure relief valve.

Install oil pump in housing and tighten mounting cap screws evenly. Carefully insert drive shaft and gear (12) into housing and through pump. Install gear (18), apply a small amount of "Loctite 270" to threads, then install and tighten locknut (19) to 85 N•m (63 ft.-lbs.) torque. Check end play of drive shaft after tightening nut. Drive shaft should rotate freely.

Lubricate bushings (1), then position drive shaft (12) and time balance weights as follows. Turn drive shaft until timing holes (A and B—Fig. 75) are aligned exactly as shown in Fig. 75, then install balance shafts (2 and 3) with flats aligned with each other and parallel to the top surface of the balancer frame. Install cover (6—Fig. 74), tightening retaining screws to 54 N•m (40 ft.-lbs.) torque, then install gasket (5) and oil transfer plate (4), tightening retaining screws to 32.5 N•m (24 ft.-lbs.) torque. Screws retaining cover (15) should be tightened to 10 N•m (7.5 ft.-lbs.) torque. Apply "Loctite 270" to threads of

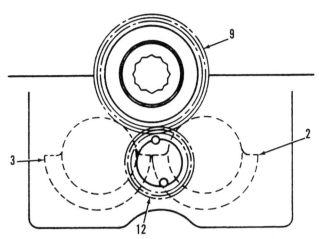

Fig. 76—Drawing of balancer assembly showing how flat surface of balance weights (2 and 3) should be aligned and parallel to mounting surface when assembling to engine block. Idler gear is shown at (9) and balancer and oil pump drive shaft gear at (12).

retaining screw, then install retainer (11), bearing (10), idler gear (9) and thrust washer (8). Tighten idler gear retaining screw to 95 N·m (70 ft.-lbs.) torque. Refer to Fig. 76 and paragraph 59 when installing.

CRANKSHAFT REAR OIL SEAL

All Models

61. The lip type rear oil seal is contained in a one-piece seal retainer attached to the rear face of the engine block. The seal and retainer can be removed after splitting tractor between the engine and transmission as outlined in paragraph 93, then removing the clutch and flywheel. Unbolt and remove seal retainer, remove old seal from retainer, then clean retainer, engine block and crankshaft.

Inspect sealing surface of crankshaft for wear or damage. If flange is grooved, new seal may be pressed slightly further into retainer so that seal will contact an unworn area of crankshaft flange. Original location of seal is with seal approximately 2.2 mm (⅛ inch) from rear of retainer. Spring loaded lip of seal is toward inside of engine. If sealing surface of crankshaft is damaged, seal contacting surface may be polished or reground to minimum diameter of 133.17 mm (5.243 inches); however, **it is important that the last 4.8 mm (³/₁₆ inch) of crankshaft remain original diameter.** The flywheel pilots on this surface of crankshaft.

Be sure that the two dowels used to locate the seal retainer are in position in block, coat gasket with sealer and position gasket on dowels. Lubricate seal and crankshaft, then carefully position seal and retainer over crankshaft, onto locating dowels against gasket and engine block. A guide (Massey-Ferguson

No. PD.145) is available to expand seal when installing over crankshaft.

FLYWHEEL

All Models

62. To remove the flywheel, first separate engine from transmission housing as outlined in paragraph 93 and remove the clutch. The flywheel is secured to the crankshaft flange by six evenly spaced cap screws. To properly time the flywheel to engine during installation, align unused hole in flywheel with untapped hole in crankshaft flange.

> CAUTION: Flywheel is only lightly piloted to crankshaft. Use caution when unbolting to prevent flywheel from falling and possibly causing injury.

The starter ring gear can be renewed after the flywheel is removed. Heat new ring gear to 245° C (475° F) before installing in position. Heat ring gear evenly and use caution to avoid overheating the gear. Heat treatment of gear could be disturbed by improper or excessive heating. Install gear on flywheel with beveled teeth toward front. Allow gear to air cool. Refer to Fig. 110 and paragraph 95 or paragraph 97 if clutch surface of flywheel needs resurfacing.

Reinstall flywheel and tighten retaining cap screws to 100 N·m (74 ft.-lbs.) torque. Check flywheel runout with a dial indicator after flywheel is installed. Runout measurement at machined outer surface should not exceed 0.30 mm (0.012 inch). Runout measured at clutch surface must not exceed 0.025 mm (0.001 inch) for each 25 mm (1 inch) from flywheel center line to the point of dial indicator measurement. If runout is excessive, remove flywheel and check for burrs or foreign material.

OIL PAN

All Models

63. The cast iron oil pan serves as part of the tractor frame and as attaching point for the tractor front support. To remove the oil pan, first drain engine oil and refer to paragraph 34 to split the front end away from the engine. Support oil pan, remove the attaching screws, then lower oil pan away from the engine.

Reinstall oil pan by reversing removal procedures.

OIL PUMP

All Models

64. The gerotor type oil pump is mounted in the center of the engine balancer housing as shown at

(16—Fig. 74) and is driven by the balancer drive shaft (12). To remove the oil pump, first remove the oil pan and engine balancer as outlined in paragraph 59. Remove nut (19), gear (18) and drive shaft (12). Remove pump mounting screws and lift pump from balancer.

Remove end cover from oil pump body and remove inner and outer rotors. Check pump body and rotors for wear or other damage. Individual parts of the pump are not available and complete assembly should be renewed.

When installing pump, tighten mounting screws to 30 N·m (22 ft.-lbs.) torque. Install drive shaft (12) and gear (18), aligning marks on gears and flats on balance weights as shown in Fig. 75 and described in paragraph 60. Reinstall and time balancer as described in paragraph 59.

RELIEF VALVE

All Models

65. The plunger type oil relief valve is located as shown at (14—Fig. 74). The relief valve is set to open at 414 kPa (60 psi.). Plunger is 37.48 mm (1.476 inches) long and original diameter is 15.95-15.98 mm (0.6279-0.6291 inch). Plunger should have 0.02-0.08 mm (0.0007-0.003 inch) diametral clearance in bore. Spring for A4.236 and A4.248S, naturally aspirated engines has 15 coils and should exert 2.54-2.94 kg (5.62-6.50 lbs.) when compressed to fitted length of 42.66 mm (1.68 inches). Spring for AT4.236, turbocharged engine has 13.5 coils and should exert 3.47-3.87 kg (7.64-8.54 lbs.) when compressed to fitted length of 42.66 mm (1.68 inches).

COOLING SYSTEM

All Models

66. RADIATOR. Remove the precleaner, hood and both lower panels. Disconnect oil cooler supply and return tubes from models so equipped. Drain coolant and disconnect upper and lower radiator hoses from all models. Unbolt fan shroud from radiator and slide shroud toward rear over the fan and coolant pump. Remove the four screws attaching the radiator, then lift the radiator up from between fuel tank and fan shroud.

To reinstall, reverse removal procedure. Coolant capacity is approximately 15.1 L (16 qts.) for naturally aspirated models without cab. Coolant capacity is approximately 15.5 L (16.4 qts.) for turbocharged models without cab. Capacity is increased for models with cab heater. Radiator cap is 0.75 bar (10 psi) rating.

67. THERMOSTAT. The thermostat (2—Fig. 77) is contained under the coolant outlet (1). Thermostat can be removed after draining coolant to below level of thermostat, detaching upper radiator hose, then unbolting thermostat housing. The thermostat should begin to open at approximately 77-85° C (170-185° F) and be fully open at 92-98° C (198-208° F).

68. COOLANT (WATER) PUMP. To remove the coolant pump, first drain the cooling system, then remove the radiator as outlined in paragraph 66. Remove the radiator shroud, loosen drive belt, then unbolt and remove the fan and belt. Disconnect hoses from pump, remove pump mounting stud nuts and screws, then remove pump. Note that some of the retaining stud nuts are behind pulley.

To disassemble pump, first remove nut (4—Fig. 77) and pulley (5). Remove internal snap ring (6), then press shaft (10) and bearings (7) toward front, out of impeller (13) and housing (3). Remainder of disassembly will be evident.

To reassemble, reverse disassembly, observe the following. Press bearings (7) and spacer (8) onto shaft (10) with shielded side of bearings away from each other. Pack bearings, then half fill the space between bearings with high temperature grease. Press the complete bearing/shaft assembly into front of housing until it bottoms. Support pulley end of pump and press impeller (13) onto shaft until clearance between

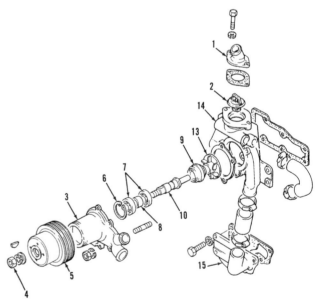

Fig. 77—Exploded view of typical coolant pump and associated parts.

1. Water outlet	
2. Thermostat	8. Spacer
3. Pump housing	9. Seal
4. Nut	10. Shaft
5. Pulley	13. Impeller
6. Snap ring	14. Housing
7. Bearings	15. Water connection

pump body and front of impeller blades is 0.28-0.89 mm (0.011-0.035 inch). Make certain that shaft rotates smoothly.

Reinstall coolant pump by reversing the removal procedure. Adjust fan belt tension to provide 10 mm (3/8 inch) deflection midway between pulleys.

TURBOCHARGER

OPERATION

390T and 398 Models

69. The Perkins AT4.236 engines used in 390T and 398 models are equipped with exhaust driven turbochargers. Garrett/AiResearch T31 turbochargers are used.

The exhaust driven turbocharger supplies air to the inlet manifold at above atmospheric pressure. The additional air entering the combustion chamber permits an increase in the amount of fuel that can be injected and burned, increasing the engine's power. Using the engine exhaust to power the compressor increases the engine's flexibility, enabling it to perform with the economy of a smaller engine on light loads yet permitting a substantial horsepower increase at full load. Horsepower loss because of altitude or atmospheric changes is also greatly reduced.

The turbocharger contains a rotating shaft with an exhaust turbine wheel on one end and a centrifugal air compressor on the other end. The exhaust turbine operates in the exhaust passage and is rotated by the exiting exhaust gasses. The compressor impeller is turned by the rotating exhaust turbine via a connecting shaft. The rotating assembly is precisely balanced and capable of rotating at very high speeds. Shaft bearings are full floating sleeve type and the unit is both lubricated and cooled by a flow of engine oil.

Do not operate the engine, even for a short time, without adequate (abundant) lubrication to the turbocharger unit. When a new or rebuilt turbocharger is first installed, if the engine has not been run for a month or more, or if a new oil filter has been installed, shut fuel OFF, then crank engine with the starter until oil pressure is at a safe level. Turn fuel ON after oil pressure reaches normal, start engine, then operate engine at slow idle speed for at least two minutes before opening throttle or putting load on engine.

Do not operate engine at wide open throttle immediately after starting engine. Allow engine to idle for a few minutes before stopping engine, to slow turbocharger speed before stopping engine.

If the engine stops suddenly while operating at heavy load and high speed, it is important to restart the engine immediately. Lack of lubrication, high rotational speed of the turbocharger and high temperature can quickly damage the turbocharger.

Maintain air filters and connections in good condition. Check system regularly and clean, renew or repair inlet air system as needed. The increased air flow of turbocharged engines may cause filters to become clogged more often than expected. Openings that permit entrance of unfiltered (dirty) air can quickly damage any engine, but damage will occur sooner on turbocharged engines because of the increased volume of air used.

Make sure exhaust pipe is covered and the air filter is connected when transporting. If the exhaust is equipped with a weathercap, tape the cap closed. If not, tape the pipe closed. If openings are not covered, air passing the open pipe may cause turbocharger to spin without adequate lubrication and damage bearings and other parts.

SERVICE

390T and 398 Models

70. REMOVE AND REINSTALL. To remove the turbocharger, first remove the tractor hood and muffler. Disconnect oil lines and air connections from turbocharger. Unbolt turbocharger from the exhaust manifold, then lift unit from tractor.

When installing, attach turbocharger to exhaust manifold using a new gasket. Attach the oil drain pipe to the turbocharger, pour 110-114 ml (4-5 fl.-oz.) of engine oil through the oil inlet port and turn turbocharger by hand to circulate the oil. Attach the oil inlet pipe, but do not tighten the screws. Operate the engine stop control while cranking the engine with starter until oil flows from the loosened inlet pipe connection, then tighten the oil supply connection. Attach air inlet connections, start engine and check for oil leaks. Complete assembly by reversing removal procedure.

71. INSPECTION AND OVERHAUL. Exchange turbocharger units are available for service, or a qualified technician can overhaul the unit if parts are available.

The Garrett/AiResearch T31 turbocharger (Fig. 78) used is equipped with a waste gate (W) to limit compressor pressure to 82 kPa (11.8 psi) by limiting exhaust turbine speed. Parts are not available for servicing the waste gate.

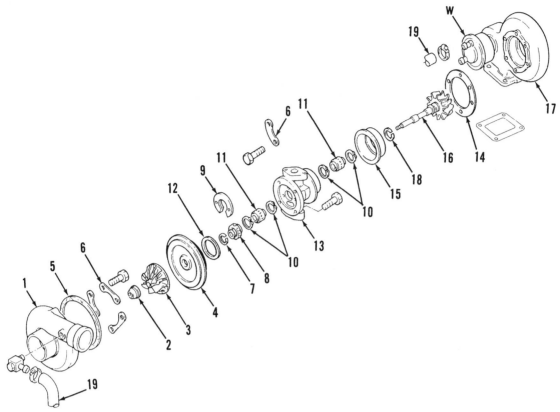

Fig. 78—Exploded view of Garrett/AiResearch T31 turbocharger used on AT4.236 engines. Waste gate unit (W) should not be disassembled.

1. Compressor housing	6. Clamp plates	15. Shroud
2. Locknut	7. Seal ring	16. Turbine shaft & wheel
3. Compressor impeller	8. Thrust collar	11. Bearings	17. Turbine housing
4. Back plate	9. Thrust bearing	12. Seal ring	18. Seal ring
5. Seal ring	10. Bearing retainers	13. Center housing	19. Hose
		14. Lock plate	

Before disassembling any turbocharger, visually inspect turbine wheel and compressor impeller blades for damage by looking through housing and end openings. Check to see if the compressor impeller or turbine blades have rubbed on the housings. Check to see if housings are wet with oil or have carbon deposits, possibly indicating engine wear or damage. Check turbocharger castings for cracks. Check lubricating oil inlet and outlet ports for carbon build up or obvious damage to bearings. Rotate shaft by hand and check for freedom of rotation, without excessive noise or play.

Mark across compressor housing (1—Fig. 78), bearing housing (13) and turbine housing (17) before disassembling, to aid alignment when reassembling. Clamp the turbocharger exhaust (mounting) flange in a vise and remove cap screws, lock plates and retaining plates, then remove the compressor housing (1). Remove cap screws, lock plates and retaining plates from other side, then lift turbocharger from the turbine housing (17).

CAUTION: Do not rest weight of any parts on impeller or turbine blades. Weight of only the turbocharger is enough to damage the blades.

Use an approved holding fixture to hold exhaust turbine and shaft (16) while removing locknut (2) with a "T" handle wrench. **Turbine shaft may be easily bent if nut is removed without using "T" handle or double universal joint.** Lift compressor impeller (3) from shaft, then withdraw exhaust turbine and shaft (16) from center housing (13) and bearings (11). Shroud (15) is not retained and will fall free when shaft is removed. Unbolt and remove back plate (4) and outside snap rings (10), then carefully withdraw bearings (11). Be careful not to damage bearings (11) or surface of center housing (13) when removing retaining snap rings (10) or bearings. The two center snap rings do not need to be removed unless damaged or unseated. Always install new snap rings if removed.

Clean all parts in cleaning solution that is not harmful to aluminum. A bristle (not wire) brush and plastic or wood scraper can be used to remove softened deposits. Use extreme caution, when cleaning, to prevent parts from being nicked, scratched or bent. Parts may be cleaned using a low pressure, dry, bead blast. Pressure must not exceed 280 kPa (40 psi). Make sure all parts are clean, smooth and free from all deposits. Drilled passages should be cleaned with compressed air.

Inspect bearing bores in center housing (13) for scored surfaces, out-of-round or excessive wear. Make sure that center housing is not grooved where seal (18) rides. Compressor impeller (3) must not show evidence of rubbing against either compressor housing (1) or back plate (4). Make certain that impeller blades are not bent, chipped, cracked or eroded. Oil passages in thrust collar (8) must be clean and thrust faces must not be warped or scored. None of the ring groove shoulders should have step wear. Inspect shroud (15) for evidence of rubbing. Turbine wheel (16) should not show evidence of rubbing and vanes should not be bent, cracked, nicked or eroded. Turbine wheel shaft should not show signs of scoring, scratching or overheating.

Always install new snap rings (10) if removed and check shaft end play and radial clearance when assembling. Position shroud (15) on turbine shaft (16) and install seal ring (18) in groove. Apply a light even coat of engine oil to shaft journals, compress seal (18) with a strong thin tool, such as a dental pick, and slide shaft and seal into center housing. Install new seal ring (7) in groove of thrust collar (8), then install thrust bearing (9) so that smooth side of bearing is toward seal ring end of collar. Install thrust bearing and collar assembly over shaft, making certain that pins in center housing engage holes in thrust bearing. Install new seal ring (12) and install back plate making certain that seal ring (7) is not damaged. Seal ring will be less likely to break if open end of seal ring is installed in back plate first. Install lock plates, retaining plates and screws to retain back plate. Install compressor impeller (3) on shaft, coat threads with engine oil and lock with nut (2). Tighten nut (2) to 2.26 N·m (20 in.-lbs.) torque, then continue to tighten nut using a "T" handle wrench until shaft length is increased 0.14-0.16 mm (0.0055-0.0065 inch). Another method of tightening nut (2) is to first tighten to 2.26 N·m (20 in.-lbs.), then turn nut an additional 110°. When nut is tight, shaft end play should be 0.03-0.10 mm (0.001-0.004 inch). Install new seal ring (5) and align compressor housing (1) and center housing (13). Alignment is easier if parts (1, 13 and 17) were marked before disassembly. Screws attaching compressor housing (1) to center housing should be tightened to 14.09 N·m (130 in.-lbs.) torque. Install exhaust housing (17) and tighten retaining screws to 14.09 N·m (130 in.-lbs.) torque. Push ends of compressor shaft while rotating and check for freedom of movement. Add engine oil to center housing and cover openings if unit is not immediately installed on tractor.

DIESEL FUEL SYSTEM

The diesel fuel system consists of three basic units; fuel tank, injection pump and injection nozzles. A transfer pump is included to move the fuel from the tank to the injection pump and filters are located in the system to make sure that the fuel is clean before entering the fuel injection pump.

Cleanliness is of the utmost importance and only approved fuel that is absolutely clean and free from foreign material should be added to the tractor fuel tank. Extra precaution should be taken to make sure that no water enters the fuel storage tanks. Extra care in daily maintenance will extend the service life and reduce costly repairs.

Extremely high precision standards are necessary in the manufacture of and service to diesel fuel injection components. Avoid nicks and burrs on all of the working parts.

FILTERS

All Models

72. MAINTENANCE. The fuel filter head is fitted with either one or two renewable type elements. Filters are equipped with water drain cocks (13—Fig. 79) at the bottom of filter. Drain cocks should be opened every few days or whenever water is visible in the sediment bowl (9). Operate lever on fuel transfer pump to refill filter with fuel. If cam for transfer pump is at point of maximum lift, hand lever may not operate.

Fuel filter (7) should be removed and new filter installed at least every 500 hours of operation. To renew the filter, turn off fuel supply at tank or next to fuel transfer pump. Clean outside of the filter and drain filter by loosening the drain cock. Remove cap screw (1) in top center of the filter head and remove

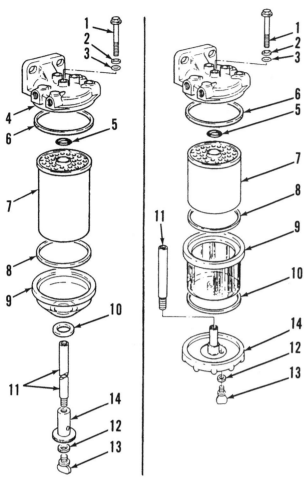

Fig. 79—Exploded view of fuel filters typical of all models. Some models may have filter head (4) that is fitted for two filter assemblies.

1. Cap screw
2. Washer
3. "O" ring
4. Filter head
5. "O" ring
6. Gasket
7. Filter element
8. Gasket
9. Bowl
10. Washer
11. Nut
12. Washer
13. Drain cock
14. Base

filter (7) and bowl (9). Clean inside of filter head and sediment bowl and make sure that all of gaskets (6, 8 and 10) and "O" rings (3 and 5) are removed. Install new filter using all new gaskets and "O" rings. Refer to paragraph 74 to bleed air from the system.

FUEL LIFT PUMP

All Models

73. The diaphragm type fuel lift (transfer) pump is mounted on the side of engine and is operated by a lobe on the engine camshaft. Several different models of fuel lift pump have been used. A repair kit containing a diaphragm, check valves and gaskets is available for servicing most pump models.

BLEEDING

All Models

74. Air must be bled from the system whenever the fuel tank has run dry, fuel filter has been renewed or any of the fuel lines have been disconnected. To bleed the system, make sure that tank shut-off is open and all fuel line connections are tight. Loosen plug or "thermostart" fuel line from top of filter head at location (B—Fig. 80). Operate priming lever of fuel lift (transfer) pump until fuel without bubbles flows from loosened plug or fitting, then tighten the plug or fitting. If cam for transfer pump is at point of maximum lift, hand lever may not operate. To correct condition, rotate engine crankshaft (and camshaft) until pump operates properly.

Move fuel shut-off control to RUN position and loosen the vent screws (1 and 2—Fig. 81) on side of pump. Actuate priming lever of fuel lift (transfer) pump until fuel free of air bubbles flows from loosened screws. Tighten lower screw (1) first, then upper screw (2). The tractor will run with air in governor housing, but bleeding at screw (2) will remove trapped air which may cause rusting. Operate priming lever and loosen fitting (T—Fig. 82) at "thermostart" unit. Tighten connection when fuel flows from fitting. Continue to operate priming lever for several strokes after tightening fitting to expel air remaining in bleed back lines. Loosen connections (N—Fig. 82), push stop control IN and move throttle to wide open. Crank engine until fuel without bubbles flows from loosened connections (N), then tighten fittings and start engine.

INJECTOR NOZZLES

CAUTION: Fuel leaves the injection nozzle with sufficient force to penetrate the skin. When testing, stay clear of the nozzle spray.

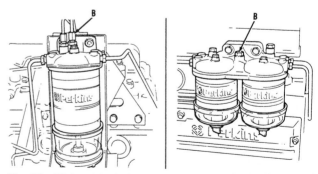

Fig. 80—View of fuel filter showing location of bleed point (B). Fitting of filter shown at left is for line to "thermostart" unit.

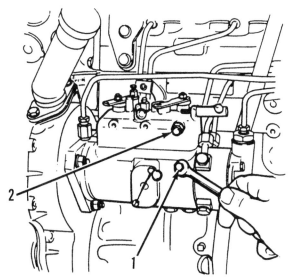

Fig. 81—View of typical fuel injection pump showing bleed plug locations (1 and 2).

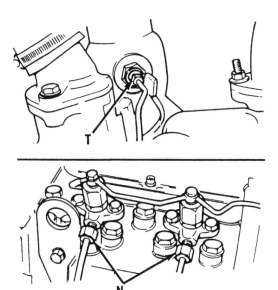

Fig. 82—View of typical high pressure connections (N) at injection nozzles and connection at "thermostart" unit (T). Refer to text for bleeding instructions.

All Models

75. LOCATING A FAULTY INJECTOR. If the engine does not run properly and a faulty injector is suspected, locate the faulty injector as follows: With the engine running, loosen the high pressure connection at each injector in turn, thereby allowing fuel to escape at the union rather then enter the injector. As in checking for malfunctioning spark plugs in a spark ignition engine, the faulty injector is the one which least affects the running of the engine when its line is loosened.

76. INJECTOR TESTING. A complete job of testing and adjusting fuel injectors requires the use of special test equipment. If injector is suspected of faulty operation, remove the injector as outlined in paragraph 81 and test for opening pressure, spray pattern, seat leakage and leak-back.

Operate the tester until oil flows, then connect the injector to the tester. Close tester valve to shut-off the passage to the gauge and operate the tester lever to be sure that injector is in operating condition and that nozzle is not plugged. If oil does not spray from all of the holes in nozzle, if test lever is hard to operate or if other obvious defects are noted, remove the injector from the tester and service as outlined in paragraph 82. If the injector operates without undue pressure on the lever and fuel is sprayed from all of the holes in nozzle, proceed with the following tests.

77. OPENING PRESSURE. Open the shut-off valve to the gauge and operate the tester lever slowly. Note the pressure indicated on the gauge at which the nozzle spray occurs. Refer to the following specifications which lists correct opening pressure and setting pressure for new injector or serviced injector with new spring.

Tractor Model	Perkins Model	Engine Build Code
362	**A4.236**	**LD31234**

Injector code – FY
Opening Pressure . . . . 17,100 kPa
(2480 psi)
Setting Pressure (new) . . . . 17,500 kPa
(2538 psi)
Nozzle. BDLL150S6705
Body BKBL67S5151

365	**A4.236**	**LD31190**

Injector code – FY
Opening Pressure . . . . 17,100 kPa
(2480 psi)
Setting Pressure (new) . . . . 17,500 kPa
(2538 psi)
Nozzle. BDLL150S6705
Body BKBL67S5151

375	**A4.236**	**LD31140**

Injector code – FY
Opening Pressure . . . . 17,100 kPa
(2480 psi)
Setting Pressure (new) . . . . 17,500 kPa
(2538 psi)
Nozzle. BDLL150S6705
Body BKBL67S5151

383 A4.248S LF31141
Injector code – FW
Opening Pressure . . . . 17,100 kPa
(2480 psi)
Setting Pressure (new) . . . . 17,500 kPa
(2538 psi)
Nozzle BDLL150S6600
Body BKJBL67S5299

390 A4.248S LF31141
Injector code – FW
Opening Pressure . . . . 17,100 kPa
(2480 psi)
Setting Pressure (new) . . . . 17,500 kPa
(2538 psi)
Nozzle BDLL150S6600
Body BKJBL67S5299

390T AT4.236 LJ31142
Injector code – HL
Opening Pressure . . . . 23,200 kPa
(3365 psi)
Setting Pressure (new) . . . . 23,200 kPa
(3365 psi)
Nozzle JB6801027
Body LRB67014

398 AT4.236 LJ31142
Injector code – HL
Opening Pressure . . . . 23,200 kPa
(3365 psi)
Setting Pressure (new) . . . . 23,200 kPa
(3365 psi)
Nozzle JB6801027
Body LRB67014

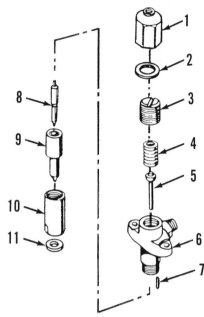

Fig. 83—Exploded view of C.A.V. injector nozzle and holder assembly typical of type used on some models.

1. Cap nut
2. Gasket
3. Adjusting screw
4. Spring
5. Spindle
6. Nozzle holder
7. Dowel pin
8. Nozzle valve
9. Nozzle body
10. Nozzle nut
11. Copper washer

cone approximately 9.5 mm (⅜ inch) from the nozzle tip. If the spray pattern does not meet these conditions, remove the injector from the tester and overhaul. If the spray pattern is satisfactory, proceed with seat leakage test as outlined in the following paragraph.

If the gauge pressure is not within limits, disassemble as required to turn the adjusting screw or change the thickness of shims as required to bring the opening pressure within specified limits. If opening pressure is erratic or cannot be properly adjusted, remove the injector from the tester and overhaul injector. Some injectors are different than type shown in Fig. 83 and are not equipped with threaded adjuster (3). Shims are used to adjust the opening pressure of some models. After opening pressure is correct, use a new copper washer and install cap nut. Recheck opening pressure after tightening cap nut to 70 N·m (50 ft.-lbs.) torque to be sure that opening pressure has not changed. If the opening pressure is within limits, check the spray pattern as outlined in the following paragraph.

78. SPRAY PATTERN. Operate the tester lever slowly and observe the nozzle spray pattern. All of the sprays must be similar and evenly spaced in nearly a horizontal plane. Each spray must be well atomized and should spread to a 76 mm (3 inches) diameter

79. SEAT LEAKAGE. Close the valve to shut-off pressure to the tester gauge and operate the tester lever quickly for several strokes. Wipe the nozzle tip dry with clean blotting paper, open the valve to the tester gauge and press the lever down slowly to bring the gauge pressure to 1035 kPa (150 psi) below the nozzle opening pressure and hold this pressure for 10 seconds. Touch a piece of blotting paper to the nozzle tip. The resulting oil blot should not be larger than 12.7 mm (½ inch) in diameter. If the nozzle tip drips oil or if blot is excessively large, the injector should be overhauled. If nozzle seat leakage is not excessive, proceed with nozzle leak-back test as outlined in the following paragraph.

80. NOZZLE LEAK-BACK. Operate the tester lever to bring gauge pressure to approximately 15,860 kPa (2300 psi), release the lever and note the time that it takes for the gauge to drop from 15,170 kPa (2200 psi) to 10,345 kPa (1500 psi). If the time is less than 5 seconds, the nozzle is worn or there are dirt particles lodged in the nozzle. If the time to drop the pressure is too long (40 seconds or more), the needle may be too tight in the bore.

> NOTE: A leaking tester connection, a leaking check valve in tester or leaking test gauge will also indicate excessively fast leak-back. If injectors consistently show excessively fast leak-back, the tester should be suspected as faulty rather than all injectors.

81. REMOVE AND REINSTALL. Carefully clean all dirt and other foreign material from lines, connectors, injectors and cylinder head area around injectors before removing the injectors. Disconnect injector leak-off line at each injector and at fuel return line. Disconnect the injector line at pump and at injector. Cover all openings and lines to prevent the entrance of dirt. Remove the two nuts or screws retaining the injector and carefully remove the injector from the cylinder head.

Make sure injector seats in cylinder head are clean and free from any carbon deposit before reinstalling injectors. Install a new copper washer in seat and a new dust sealing washer around the body of injector. Insert injector in cylinder head bore, install retaining washers and nuts or screws. Tighten retaining screws or nuts evenly and alternately to 16 N·m (12 ft.-lbs.) torque. Install a new gasket below and above each leak-off banjo fitting, then tighten the banjo bolt to 8 N·m (72 in.-lbs.) torque. Reconnect the leak-off line to the return line. Check fuel injector connections to be sure that they are clean and properly aligned with fittings at injector and at pump. Reconnect injector pressure lines, tightening to 27 N·m (20 ft.-lbs.) torque.

It is suggested that the fuel filter be serviced as described in paragraph 72 each time injectors are serviced. Refer to paragraph 74 for bleeding the fuel system. Since lines to injectors are opened, it will be necessary to crank engine with fitting at injectors loosened to fill the lines. Start and run engine to be sure that injectors are properly sealed and that injector pressure line and leak-off line connections are not leaking.

82. OVERHAUL. Do not attempt to overhaul diesel injectors unless complete and proper equipment is available. Refer to Fig. 83 for exploded view of one type of injector unit.

Secure the injector holding fixture in a vise and locate the injector in the holding fixture. Never clamp the injector body in a vise. Remove the cap nut and back-off the adjusting screw of models so equipped. Lift the upper spring disc, injector spring and lower spring seat from injector. Remove the nozzle retaining nut using an appropriate special socket, then remove nozzle and valve. The nozzle and valve are a lapped fit to each other and parts from one must never be interchanged with parts from another matched unit. Place all parts in clean fuel oil or calibrating fluid as they are disassembled. Clean the injector unit as follows: Soften hard carbon deposits by soaking in a suitable carbon solvent, then use a soft wire (brass) brush to remove carbon from the exterior. Rinse the nozzle and needle immediately after cleaning to prevent carbon solvent from corroding the highly finished surfaces. The pressure chamber of nozzle can be cleaned with a special reamer. Clean spray holes (Fig. 84) with proper size of wire probe held in a pin vise as shown in Fig. 85. To reduce the chance of breakage, wire should protrude from vise only far enough to pass through spray holes. Rotate pin vise, but do not apply undue pressure.

Valve seat in nozzle is cleaned by inserting a special seat scraper into nozzle, then rotating the scraper.

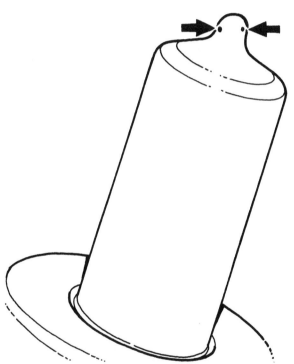

Fig. 84—Holes (arrows) in the injector nozzle are not located an equal distance from the nozzle tip.

Fig. 85—Clean spray holes with a pin vise and wire probe as shown.

The annular groove can be cleaned with another special hooked scraper.

After cleaning, back-flush the nozzle using a back-flush adapter attached to nozzle tester. Rotate the nozzle while operating the lever to flush nozzle.

Seat in nozzle can be polished using a small amount of tallow on the end of a polishing stick. Be sure that carbon is thoroughly flushed from nozzle before polishing seat.

If the leak-back test was longer than 40 seconds as tested in paragraph 80, or if needle is sticking in bore of nozzle, the needle should be lapped to the nozzle. Use special polishing lapping compound such as Bacharach 66-0655. Place the small diameter of nozzle in a drill chuck and rotate at not more than 450 rpm. Apply a small amount of polishing compound on the barrel of valve needle taking care not to allow any compound on tip or beveled portion. Insert the valve needle into the slowly rotating nozzle. Work the needle in and out several times taking care not to put any pressure against seat. Withdraw needle, remove nozzle and back-flush the nozzle and needle.

NOTE: Do not lap valve for longer than 5 seconds at a time and allow parts to cool between lappings.

Assemble nozzle and needle while still wet from cleaning. Valve needle should slide easily in nozzle. If any sticking is noted, be sure that parts are clean and temperature has normalized.

Rinse all parts in clean fuel oil or calibrating fluid before reassembling, then assemble while still wet. Position the nozzle and needle valve on injector body and be sure that dowel pins in body are correctly located in nozzle (Fig. 86). Install the nozzle retaining nut and tighten nut to 70 N·m (50 ft.-lbs.) torque. Install spindle, washer, spring and pressure adjusting screw. Connect the injector to tester and adjust opening pressure as outlined in paragraph 77. Some injectors are different than type shown in Fig. 83 and are not equipped with threaded adjuster (3). Shims are used to adjust the opening pressure of some

models. After opening pressure is correct, use a new copper washer and install cap nut. Recheck opening pressure after tightening cap nut to 70 N·m (50 ft.-lbs.) torque to be sure that opening pressure has not changed.

Retest injector as outlined in paragraphs 76 through 80. Install new nozzle and needle assembly or complete injector assembly if still faulty. Nozzles should be thoroughly flushed with calibrating fluid prior to storage.

FUEL INJECTION PUMP

All Models

83. TIMING. Refer to paragraph 49 for installation and timing of the injection pump drive gear. The pump is internally timed and except for adjusting to compensate for timing gear backlash, timing can be considered correct if mark on pump mounting flange is aligned with mark on the engine as shown in Fig. 87. Refer to paragraph 86 for specific application and timing specifications.

84. REMOVE AND REINSTALL. Thoroughly clean the pump, lines and connections in the area around pump. Remove the lines from the pump to the injectors, disconnect the fuel inlet and outlet (return) lines and immediately cover all openings to prevent the entrance of dirt. Disconnect the throttle control rod and the fuel shut-off cable.

The fuel injection pump is mounted horizontally. To remove the pump, remove cover plate located in front of injection pump from the timing gear cover. Remove the three screws attaching the injection pump drive gear to the pump shaft. Unbolt the injection pump from the engine front plate and withdraw pump. **Be**

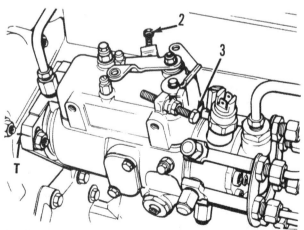

Fig. 87—Timing marks (T) on pump mounting flange and mounting pad should be aligned as shown when installing pumps that are mounted horizontally. Low idle speed set screw is shown at (2) and maximum speed set screw is located at (3).

Fig. 86—Be sure that lapped surfaces are clean and that locating dowels are perfectly aligned when nozzle body is reinstalled.

careful not to drop locating pin or drive key when removing pump. It may be necessary to use a puller to push the pump drive shaft from the gear. The pump drive gear will remain in the timing gear housing and will not change the gear timing. The gear can not be removed unless cover is removed; however, the engine crankshaft should not be turned with the pump removed.

To reinstall the fuel injection pump, reverse the removal procedures. Align the scribed mark on pump body mounting flange with mark on engine timing gear housing as shown in Fig. 87 and tighten pump retaining nuts to 24 N·m (18 ft.-lbs.) torque. Tighten the three screws attaching the injection pump gear to the injection pump shaft to 38 N·m (28 ft.-lbs.) torque. Connect fuel lines, throttle control rod and fuel shut-off cable, then bleed the fuel system as outlined in paragraph 74.

INJECTION PUMP DRIVE GEAR

All Models

85. To remove the injection pump drive gear, first remove the timing gear cover or housing as outlined in paragraph 44. Remove the rocker arm assembly and the camshaft drive (idler) gear. Damage could result if either camshaft or crankshaft is turned independently from the other unless the rocker arms are removed. Remove the three screws attaching gear to pump shaft, then remove the gear. It may be necessary to use a puller to pull the gear from the pump drive shaft.

When installing the gear, align dowel pin with slot in adapter hub or drive key with slot. Be sure gear timing marks are aligned as shown in Fig. 65.

INJECTION PUMP

All Models

86. The injection pump is a completely sealed unit. Service of any kind should only be attempted by trained personnel with proper equipment to test, repair and adjust CAV pumps of the type installed. Refer to the following for specific application and timing specifications.

Tractor Model	Perkins Model	Engine Build Code
362	**A4.236**	**LD31234**

Pump – Standard. . . . . . . DPA 2643C282LS/3/2310
   ISO . . . . DPA LS49L/900/3/2310
   Rotation . . . . Clockwise
   Snap ring timing mark letter C
   No. 1 Cyl. Outlet letter . . . . W
   Engine Checking Angle at TDC . . . . 284.5°

   Fuel Pump Checking Angle. . . . . 296°
   Static Timing BTDC . . . . 23°
   Piston Displacement . . . . 6.4575 mm
                           (0.2542 inch)

365	**A4.236**	**LD31190**
Before U261449S		

Pump – Standard. . . . . DPA 38313LS/2/2420
   ISO . . . . DPA LS49L/900/2/2420
   Rotation . . . . Clockwise
   Snap ring timing mark letter C
   No. 1 Cyl. Outlet letter . . . . W
   Engine Checking Angle at TDC . . . . 284.5°
   Fuel Pump Checking Angle. . . . . 296°
   Static Timing BTDC . . . . 23°
   Piston Displacement . . . . 6.4575 mm
                           (0.2542 inch)

After U261448S

Pump – Standard. . . . . . DPA 2643C282LS/3/2310
   ISO . . . . DPA LS49L/900/3/2310
   Rotation . . . . Clockwise
   Snap ring timing mark letter C
   No. 1 Cyl. Outlet letter . . . . W
   Engine Checking Angle at TDC . . . . 284.5°
   Fuel Pump Checking Angle. . . . . 296°
   Static Timing BTDC . . . . 23°
   Piston Displacement . . . . 6.4575 mm
                           (0.2542 inch)

375	**A4.236**	**LD31140**
Before U209016P		

Pump – Standard. . . . . DPA 38377LS/2/2420
   ISO . . . . DPA LS57L/900/2/2420
   Rotation . . . . Clockwise
   Snap ring timing mark letter C
   No. 1 Cyl. Outlet letter . . . . W
   Engine Checking Angle at TDC . . . . 284.5°
   Fuel Pump Checking Angle. . . . . 296°
   Static Timing BTDC . . . . 23°
   Piston Displacement . . . . 6.4575 mm
                           (0.2542 inch)

After U209015P

Pump – Standard. . . . . . DPA 2643C278LS/3/2310
   ISO . . . . DPA LS57L/900/3/2310
   Rotation . . . . Clockwise
   Snap ring timing mark letter C
   No. 1 Cyl. Outlet letter . . . . W
   Engine Checking Angle at TDC . . . . 284.5°
   Fuel Pump Checking Angle. . . . . 296°
   Static Timing BTDC . . . . 23°
   Piston Displacement . . . . 6.4575 mm
                           (0.2542 inch)

383	**A4.248S**	**LF31141**

Pump – Standard. . . . . DPA 38367XS/8/2380
   ISO . . . . DPA XS59L/800/8/2380
   Rotation . . . . Clockwise
   Snap ring timing mark letter C

No. 1 Cyl. Outlet letter . . . . W
Engine Checking Angle at TDC . . . . 281°
Fuel Pump Checking Angle. . . . . 293°
Static Timing BTDC . . . . 24°
Piston Displacement . . . . 7.0177 mm
(0.2763 inch)

390   **A4.248S**   **LF31141**
Before U226835S
Pump – Standard. DPA38367XS/8/2380
ISO . . . . DPA XS59L/800/8/2380
Rotation . . . . Clockwise
Snap ring timing mark letter C
No. 1 Cyl. Outlet letter . . . . W
Engine Checking Angle at TDC . . . . 281°
Fuel Pump Checking Angle. . . . . 293°
Static Timing BTDC . . . . 24°
Piston Displacement . . . . 7.0177 mm
(0.2763 inch)

After U226834S
Pump – Standard. . . . . . DPA 2643C279XS/6/2310
ISO . . . . DPA XS65L/800/6/2310
Rotation . . . . Clockwise
Snap ring timing mark letter C
No. 1 Cyl. Outlet letter . . . . W
Engine Checking Angle at TDC . . . . 281°
Fuel Pump Checking Angle. . . . . 293°
Static Timing BTDC . . . . 24°
Piston Displacement . . . . 7.0177 mm
(0.2763 inch)

390T   **AT4.236**   **LJ31142**
Before U219832S
Pump – Standard. . . . . . DPA 2643C188KT/5/2420
ISO . . . . DPA KT68L/1000/5/2420
Rotation . . . . Clockwise
Snap ring timing mark letter C
No. 1 Cyl. Outlet letter . . . . W
Engine Checking Angle at TDC . . . . 284°

Fuel Pump Checking Angle. . . . . 292°
Static Timing BTDC . . . . 16°
Piston Displacement . . . . 3.1602 mm
(0.1244 inch)

After U219831S
Pump – Standard. . . . . . DPA 2643C280KT/8/2310
ISO . . . . DPA KT68L/1000/8/2310
Rotation . . . . Clockwise
Snap ring timing mark letter C
No. 1 Cyl. Outlet letter . . . . W
Engine Checking Angle at TDC . . . . 284°
Fuel Pump Checking Angle. . . . . 292°
Static Timing BTDC . . . . 16°
Piston Displacement . . . . 3.1602 mm
(0.1244 inch)

398   **AT4.236**   **LJ31142**
Before U219832S
Pump – Standard. . . . . . DPA 2643C188KT/5/2420
ISO . . . . DPA KT68L/1000/5/2420
Rotation . . . . Clockwise
Snap ring timing mark letter C
No. 1 Cyl. Outlet letter . . . . W
Engine Checking Angle at TDC . . . . 284°
Fuel Pump Checking Angle. . . . . 292°
Static Timing BTDC . . . . 16°
Piston Displacement . . . . 3.1602 mm
(0.1244 inch)

After U219831S
Pump – Standard. . . . . . DPA 2643C280KT/8/2310
ISO . . . . DPA KT68L/1000/8/2310
Rotation . . . . Clockwise
Snap ring timing mark letter C
No. 1 Cyl. Outlet letter . . . . W
Engine Checking Angle at TDC . . . . 284°
Fuel Pump Checking Angle. . . . . 292°
Static Timing BTDC . . . . 16°
Piston Displacement . . . . 3.1602 mm
(0.1244 inch)

ELECTRICAL SYSTEM

87. A 12 volt, negative grounded electrical system is used on all models.

ALTERNATOR AND REGULATOR

All Models

88. Lucas or Motorola 45, 55 or 70 amp alternators have been used. All models have the regulator built into the alternator.

To disassemble the Motorola alternator, first scribe a line across frame halves (5 and 21—Fig. 88). Remove voltage regulator (25), brush set (23), insulating cover (26), then remove the four through-bolts. Carefully pry frame apart, between stator frame (15) and drive end (5). **Be careful not to damage windings of stator.** Unsolder stator and voltage regulator leads from rectifier bridge (11). Remove rectifier bridge, capacitor (13) and stator (15). Remove "O" ring (16) from slip ring end frame (21), then remove D+ terminal screw and spade connector/insulator (24) from slip ring end frame (21).

Remove pulley, fan (1), washer (2), Woodruff key (4) and spacer (3), then tap rotor shaft on block of wood to push it out of bearing in drive end frame (5). Slide counterbore spacer (8) off rotor shaft (9). Press bear-

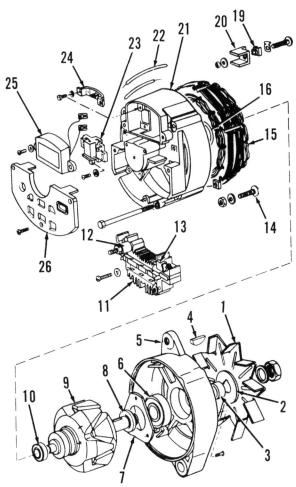

Fig. 88—Exploded view of Motorola alternator typical of type used on some models. Refer to Fig. 92 for Lucas type.

1. Fan	13. Capacitor
2. Washer	14. Ground terminal screw
3. Spacer	15. Stator
4. Woodruff key	16. "O" ring
5. Drive end frame	19. Insulator
6. Bearing	20. Insulator
7. Bearing retainer	21. Slip ring end frame
8. Counterbored spacer	22. Wire
9. Rotor	23. Brush set
10. Bearing	24. Spade connector
11. Rectifier bridge	25. Voltage regulator
12. Spacer	26. Insulating cover

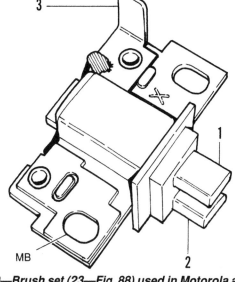

Fig. 89—Brush set (23—Fig. 88) used in Motorola alternator. Refer to text for test procedures.

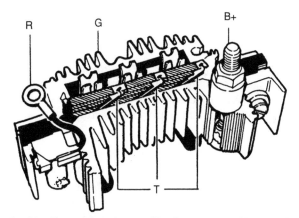

Fig. 90—Test points for rectifier bridge (11—Fig. 88). Refer to text for testing procedure.

ing (6) out of drive end frame (5), after removing bearing retainer (7). Pull bearing (10) off rotor shaft.

Original brush set (23) can be reinstalled if brushes protrude ³⁄₁₆ inch or more from bottom of holder. Brushes should not be oil soaked or cracked. Check for continuity from brush (1—Fig. 89) to terminal (3) and from brush (2) to mounting bracket (MB), using an ohmmeter. Check for open circuit between brushes (1 and 2) and from terminal (3) and mounting bracket (MB).

Rectifier bridge (11—Fig. 88) may be tested with an ohmmeter after unsoldering stator and voltage regulator leads. Connect one ohmmeter lead to ground

(G—Fig. 90) and other lead to one of the terminals (T), then note ohmmeter reading. Reverse ohmmeter leads and again note reading. A good diode will read approximately 30 ohms resistance in one test mode and continuity (near zero resistance) with test leads reversed. Repeat test on remaining terminals (T). Repeat test between (B+) and terminals (T), then repeat test between (R and T) terminals. Install new rectifier bridge if any one pair of readings is the same regardless of test lead polarity.

Check bearing surfaces of rotor shaft (9—Fig. 88) for visible wear or scoring. Examine slip ring surface for scoring or wear and rotor winding for overheating

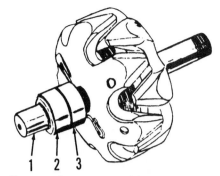

Fig. 91—Removed rotor assembly showing test points to be used when checking for grounds (shorts) and opens.

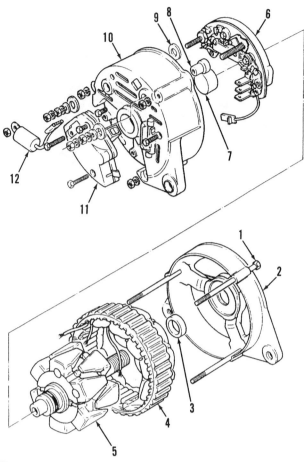

Fig. 92—Exploded view of Lucas alternator typical of type used on some models.

1. Screws	7. Bearing
2. Drive end frame	8. Insulator
3. Spacer	9. Washer
4. Stator	10. Slip ring end frame
5. Rotor	11. Regulator
6. Rectifier	12. Capacitor

or other damage. Check rotor for grounded, shorted or open circuits using an ohmmeter. Touch ohmmeter probes to points (1 and 2—Fig. 91) and to points (1 and 3). A reading near zero indicates a short to ground. Touch ohmmeter probes to slip rings (2 and 3) and observe resistance which should be 4.0-5.2 ohms. If resistance is too high, windings are open. If ohmmeter indicates less than 4.0 ohms, windings are shorted or grounded and new rotor should be installed.

Clean slip rings (2 and 3) with fine crocus cloth while spinning the rotor to avoid flat spots on slip rings.

To reassemble, reverse disassembly procedure. When reconnecting lead wires from voltage regulator (25—Fig. 88), connect yellow wire to D+ terminal and green wire to brush assembly connector.

The Lucas alternator used on some models is shown in Fig. 92. Regulator and brushes (11) can be removed without disassembling unit. Alternator can be disassembled after removing the nut, pulley and fan from front, then removing the three screws (1).

STARTER

All Models

89. Both Lucas and Perkins starters have been used. Construction is typical for both models as shown in Fig. 93 and Fig. 94.

CIRCUIT DESCRIPTION

All Models

90. Refer to Fig. 95, Fig. 96, Fig. 97, Fig. 98, Fig. 99, Fig. 100, Fig. 101, Fig. 102, Fig. 103 and Fig. 104 for wiring diagrams typical of most models. Wiring for some models may be different than shown. All models include 12 volt batteries with the negative terminal grounded. Quick disconnects are provided at all main connections. If trouble is encountered and wiring is suspected, check for loose, corroded, damaged or disconnected connections.

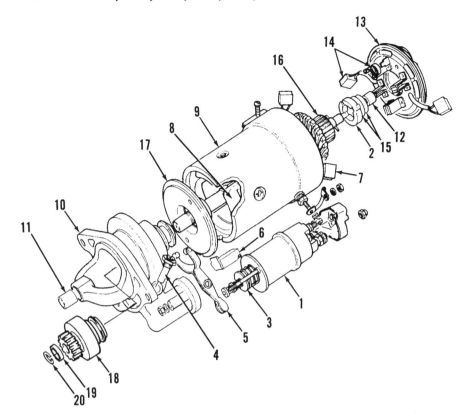

Fig. 93—Exploded view of Lucas starter used on some models.

1. Solenoid
2. Brake
3. Plunger
4. Pivot pin
5. Shift lever
6. Dust seal
7. Brush
8. Field coil
9. Field housing
10. Drive end frame
11. Bushing
12. Bushing
13. Brush holder & end cover
14. Brush & spring
15. Washers
16. Armature
17. Center plate
18. Drive pinion
19. Pinion stop collar
20. Snap ring

Fig. 94—Exploded view of Perkins starter assembly.

1. Solenoid
2. Spacer washers
3. Plunger
4. Pivot pin
5. Shift lever
6. Dust seal
7. Brush
8. Field coil
10. Drive end frame
11. Bushing
12. End cover
13. Brush holder
15. Washers
16. Armature
17. Center plate
18. Drive pinion
19. Pinion stop collar
20. Snap ring

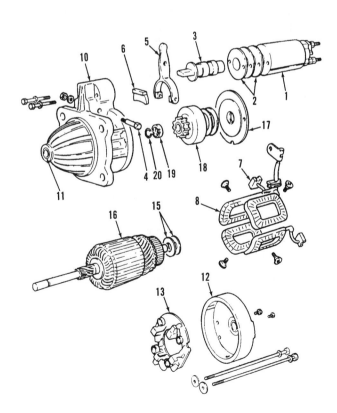

Color code and legend for Figs. 95-104.

B. Black
BU. Black/Blue
BP. Black/Purple
BR. Black/Red
BS. Black/Slate
BLG. Black/Light Green
BW. Black/White
G. Green
GB. Green/Black
GN. Green/Brown
GO. Green/Orange
GP. Green/Purple
GR. Green/Red
GU. Green/Blue
GW. Green/White
K. Pink
LG. Light Green
LGY. Light Green/Yellow
LU. Light Blue
N. Brown
NB. Brown/Black
NG. Brown/Green
NK. Brown/Pink
NR. Brown/Red
NU. Brown/Blue
O. Orange
P. Purple
PB. Purple/Black
PN. Purple/Brown
R. Red
RB. Red/Black
RU. Red/Blue
RG. Red/Green
RLG. Red/Light Green
RW. Red/White
RY. Red/Yellow
S. Grey
U. Blue
UG. Blue/Green
UK. Blue/Pink
UR. Blue/Red
UW. Blue/White
W. White
WB. White/Black
WG. White/Green
WK. White/Pink
WLG. White/Light Green
WN. White/Brown
WR. White/Red
WS. White/Grey
Y. Yellow
YU. Yellow/Blue

1. Horn
2. Headlights
3. Fuel level sender
4. Solenoid – 4WD Differential lock

5. Switch – Engine oil pressure
6. Sender – Coolant temperature
7. Switch – Air filter
8. Switch – Hydraulic filter temperature
9. Switch – Hydraulic filter pressure
10. Alternator
11. Starter motor
12. Battery
13. Thermostart unit
14. Cigar lighter
15. Switch – Stop light
16. Trailer indicator warning light
17. Direction indicator warning light
18. Switch – Direction indicator
19. Flasher unit – Direction indicator
20. Start switch
21. Main fuse box
22. Thirteen-Pin plug – Instrument panel warning lights
23. Twelve-Pin plug – Instrument panel instruments
24. Switch – Main/dip beam
25. Switch – Headlight
26. Switch – Rear work light
27. Switch – Hazard warning light
28. Warning light dimmer
29. Safety start switch – Transmission neutral
30. Horn button
31. Idle speed indicator light
32. Fuse – Main start circuit
33. Speedometer
34. Safety start switch – PTO
35. Switch – Differential lock engage
36. Front fender lights – Right
37. Rear fender lights – Right
38. Rear work light – Right
39. Trailer socket
40. License plate light
41. Switch – Parking brake
42. Switch – 4WD
43. Front fender lights – Left
44. Rear fender lights – Left
45. Ground (–) junction
46. Feed (+) junction
47. IPTO switch
48. Red quick disconnect – Behind engine
49. Blue quick disconnect – Behind engine
50. Black quick disconnect – Behind engine
51. Quick disconnect – Right clutch housing
52. Quick disconnect – Right side cover
53. Quick disconnect – Right clutch housing
54. Quick disconnect – Right side cover
55. Quick disconnect – Left seat deck
56. Quick disconnect – Right seat deck
57. Quick disconnect – Speedometer
58. Quick disconnect – Speedometer
59. Rear work light – Left
60. Front work light – Right

61. Front work light – Left
62. Beacon
63. Interior light (w/cab)
64. Front wiper motor (w/cab)
65. Rear wiper motor
66. Radio
67. Switch – Road Hi-Lo
68. Switch – Rear wiper
69. Switch – Front wiper
70. Switch – Cab heater blower
71. Switch – Beacon
72. Cab heater blower motor
73. Switch – Multi-Power
74. Left flasher
75. Quick disconnect
76. Quick disconnect
77. Right flasher
78. Switch – Front work light
79. Windshield washer pump
80. Cab fuse box
81. Multi-Power solenoid
82. Switch – Reverse horn
83. Reverse horn
84. Starter motor solenoid
85. Road warning lights
86. Switch – Road warning light
87. Flasher – Road warning light
88. Radio speaker
89. Radio antenna
90. Engine ground
91. Air conditioner relay
92. Air conditioner thermostat
93. Air conditioner compressor
94. Switches – Air conditioner cut-out
95. Quick disconnect – Air conditioner
96. Rear window washer pump
97. Fresh air blower motor
98. Fresh air blower motor resistor
99. Switch – Fresh air blower
100. Line fuse – Air conditioner
101. Fresh air blower motor relay
102. Quick disconnect – Right light
103. Quick disconnect – Left light
104. Quick disconnect – Direction indicator
105. Switch – 4WD
106. Range indicator light junction
107. Range indicator light junction
108. Range indicator light switch junction
109. Switch – Range indicator light
110. Quick disconnect – Reverse horn
111. Quick disconnect – Reverse horn switch
112. Quick disconnect – Differential lock switch
113. Engine speed sensor
114. Fuel valve
115. Quick disconnect – 4WD switch
116. Switch – Battery isolator
117. Tachometer interface unit
118. 4WD solenoid valve

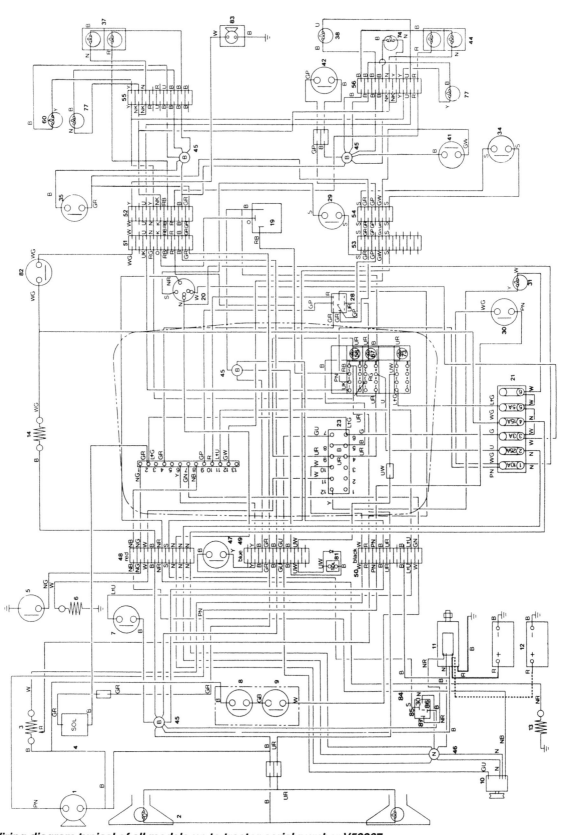

Fig. 95—Wiring diagram typical of all models up to tractor serial number V52267.

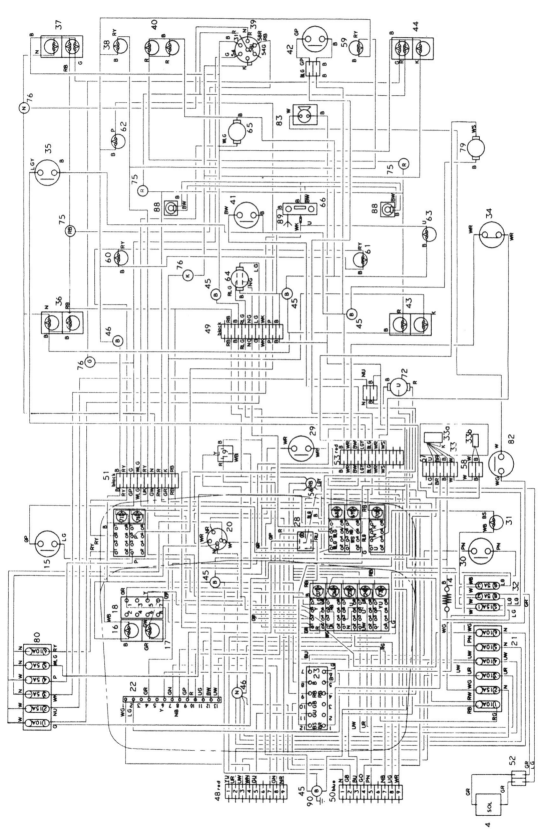

Fig. 96—Wiring diagram for models with cab from tractor serial number V52268 to P22199.

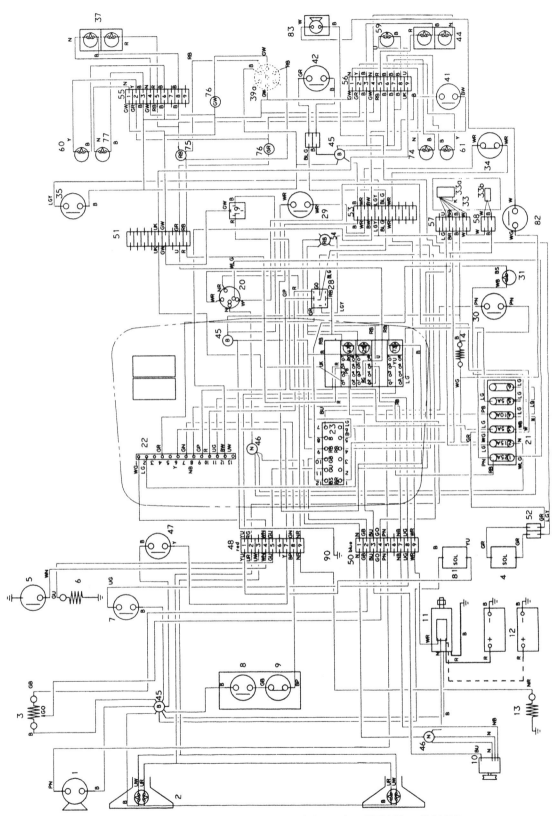

Fig. 97—Wiring diagram for models without cab from tractor serial number V52268 to R14072.

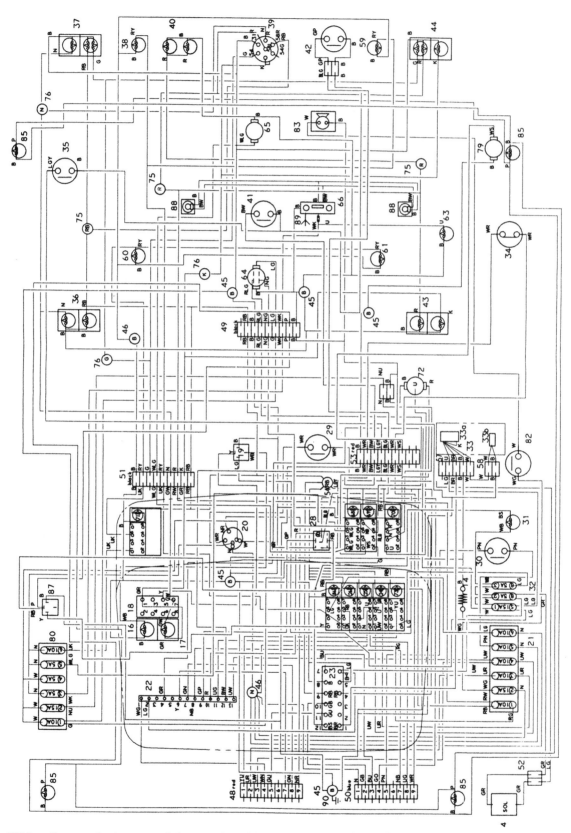

Fig. 98—Wiring diagram for tractor cab from serial number V52268 to P22199. Refer to Fig. 99 for front circuit.

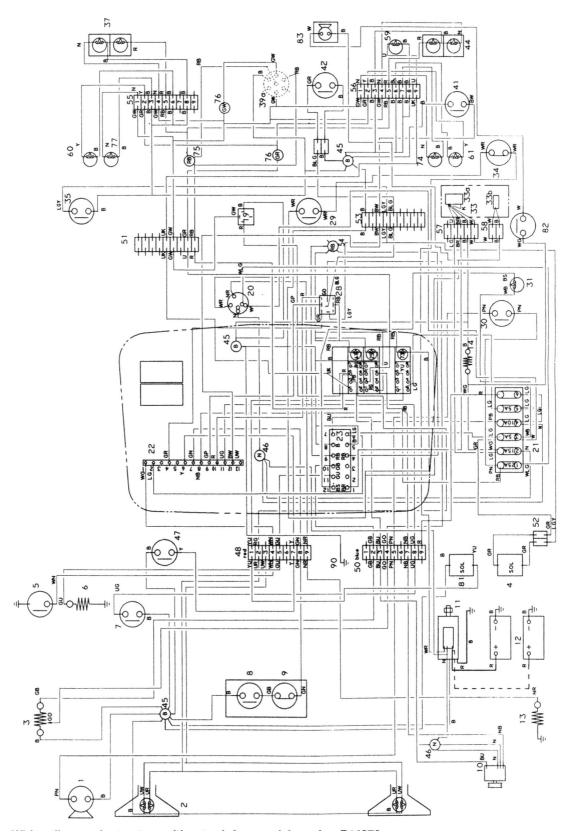

Fig. 99—Wiring diagram for tractors without cab from serial number R14073.

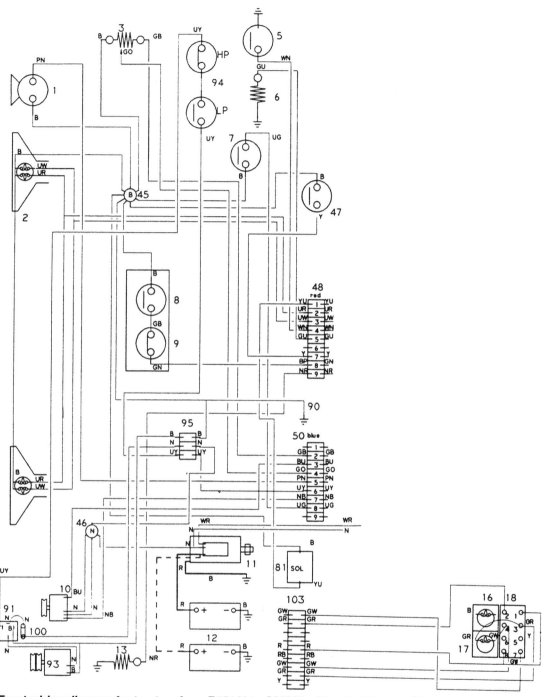

Fig. 100—Front wiring diagram for tractors from R15144 to S00000 with cab. Refer to Fig. 97 for rear circuit.

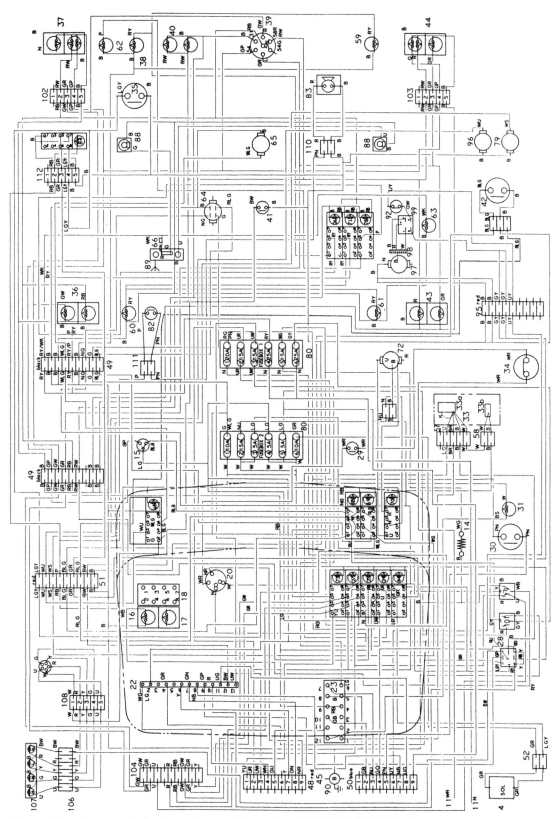

Fig. 101—Rear wiring diagram for tractors from serial number R15144 to S00000 with cab.

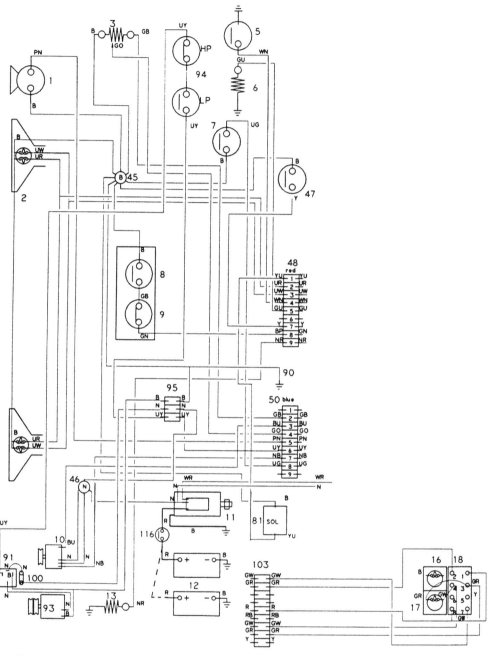

Fig. 102—Front wiring diagram for models with cab after serial number S01000. Refer to Fig. 99 for rear circuit.

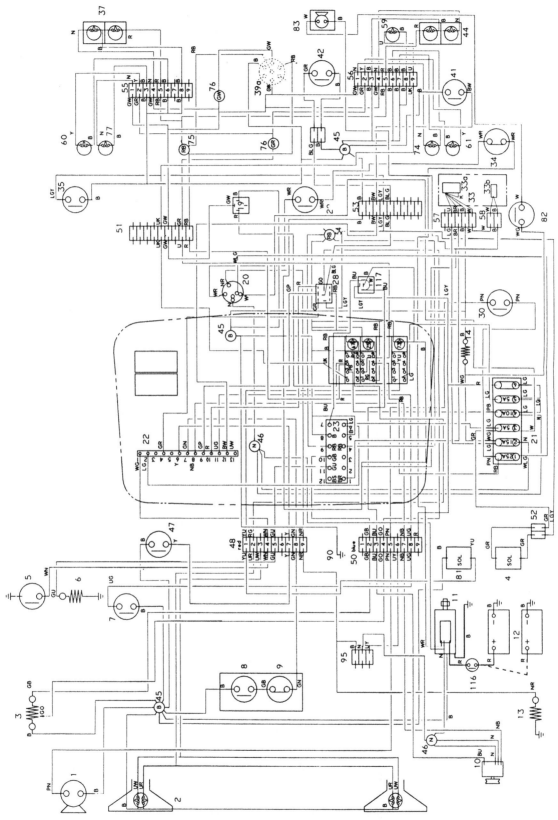

Fig. 103—Wiring diagram typical of all models after serial number S01000 without cab.

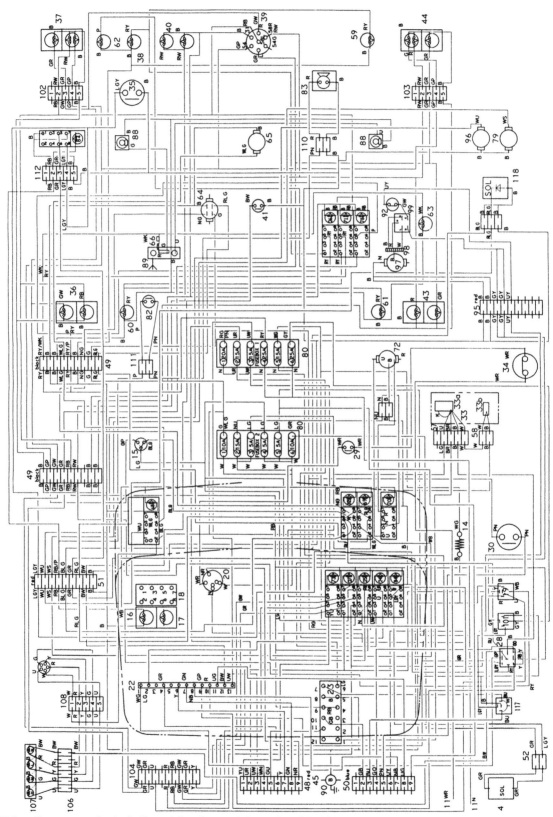

Fig. 104—Wiring diagram typical of all models after serial number S01000 with cab.

ENGINE CLUTCH

91. All models are equipped with either a dual or split torque clutch. Some models are equipped with 305 mm (12 inches) diameter clutch and others have a larger 330 mm (13 inches) diameter unit. Other differences will also be noted in the following service and adjustment procedures when necessary. The dual clutch controls engagement of drive to the engine transmission and to the power take off (pto). Split torque clutches are installed on tractors equipped with independent pto. Drive for the hydraulic pump and for the pto are taken from the splined hub of a plate attached to the clutch cover, which turns all the time that engine is running. Control of pto engagement or disengagement is by a separate, hydraulically operated clutch. Refer to the appropriate following paragraphs for service to both types of clutches.

CLUTCH LINKAGE ADJUSTMENT

All Models

92. Clutch linkage adjustment should be carefully monitored at all times, but especially during the first 50 hours of operation with new clutch linings.

On all models except 362, 365 and 375 models with split torque clutch, pedal height (C—Fig. 105) is set by the upper attachment position of the main link. If tractor is equipped with dual clutch, use lower hole (L) as shown. If equipped with split torque clutch, attach link using upper hole (U).

On 362, 365 and 375 models with split torque clutch, pedal height (C—Fig. 106) is adjusted by stop screw (S). Pedal height when released should be 120 mm (4.75 inches). Some later tractors may have fixed stop and are not adjustable.

On all models, clutch pedal free play should be measured by pressing pedal down until release lever just contacts clutch release fingers and measure distance (A—Fig. 105 or Fig. 106) pedal has traveled. On all models except 383 and 390 with dual clutch, free travel should be 20-25 mm (¾ - 1 inch). A gas assist cylinder is located at (G—Fig. 105) on 383 and 390 models with dual clutch and free travel should be 10-15 mm (⅜ - ⅝ inch). Free travel is adjusted on all models by turning nut (N—Fig. 105 or Fig. 106).

To adjust the pto clutch on models with dual clutch and live pto, remove the cover from underside of the clutch housing and turn engine until one of the three adjusters is visible as shown in Fig. 107. Clearance (C) between head of screw and clutch plate should be

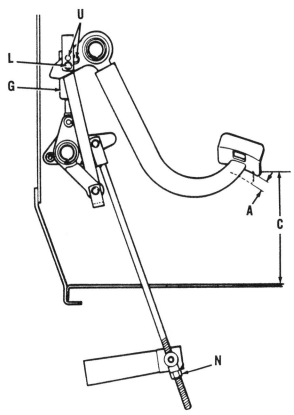

Fig. 105—View of clutch pedal and operating linkage typical of all models except 362, 365 and 375 models with split torque clutch.

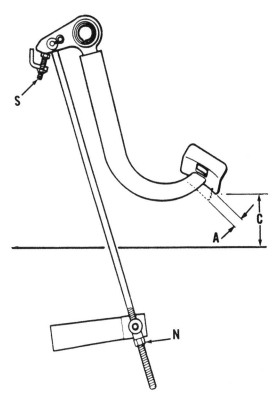

Fig. 106—View of clutch pedal and operating linkage typical of 362, 365 and 375 models with split torque clutch.

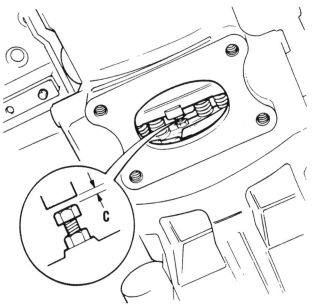

Fig. 107—View of one of three adjusters of live pto clutch. Clearance (C) should be 2 mm (0.080 inch) for all three locations.

2 mm (0.080 inch). Screw can be threaded in or out as necessary after loosening locknut. Make sure that clearance at all three adjusters is exactly alike.

ENGINE/TRANSMISSION SPLIT

All Models

93. To separate tractor between rear of engine and front of transmission, block both rear wheels to prevent rolling, remove any front mounted equipment and any front weights. Remove hood, grille and side panels. If equipped with a cab heater, drain coolant from radiator and engine block, then disconnect hoses to heater. If equipped with air conditioning, disconnect refrigerant lines. Remove gear lever trim plate and boot from models with cab. On four-wheel-drive tractors, refer to paragraph 9 and remove drive shaft and shield. On all models, disconnect battery ground and front wiring harness to the headlights, horn and fuel tank. Disconnect throttle and stop cables from the fuel injection pump, detach cables from support brackets and move cables out of the way. Disconnect hydraulic lines between rear of tractor and front that would interfere with the removal of engine and cover all openings. If necessary, drain fuel tank. Wedge between the front support and axle to prevent tipping and attach hoist to the engine. Support rear of tractor under transmission housing, then carefully remove the screws and nuts attaching the engine to transmission housing. Carefully move the engine and front axle assembly forward away from the transmission.

Reassemble by reversing the splitting procedure and observe the following. Install a M12 alignment

stud approximately 100 mm (4 inches) long in each side of the transmission housing to assist in aligning engine to transmission. Turn flywheel to align clutch plate splines with transmission and pto input shaft splines. Install retaining screws and nuts after engine and transmission housing flanges are completely together. Tighten retaining screws and nuts to 115 N·m (85 ft.-lbs.) torque. Reverse removal procedure and refer to paragraph 5 for attaching front support to engine and selecting proper shims. Check and adjust clutch linkage as outlined in paragraph 92. Refer to paragraph 9 for installing drive shaft of four-wheel-drive models.

CLUTCH ASSEMBLY

All Models With Dual Clutch

94. REMOVE AND REINSTALL. To remove the clutch assembly, first refer to paragraph 93 and separate engine from transmission housing. Install three ¼ UNC screws 2⅛ inches (54 mm) long into the three equi-spaced holes (3—Fig. 108) in clutch cover to hold. Loosen the six screws (6) retaining the clutch to flywheel, then remove screws and separate clutch from flywheel. **Be careful to avoid breathing dust which may contain asbestos. Do not use compressed air to blow dust from clutch and flywheel.**

Position main clutch disc on flywheel and center using alignment tool (MF 159B or equivalent). Make sure three screws (3) are installed and tightened to compress springs, then locate clutch assembly and install the six retaining screws (6). Tighten retaining

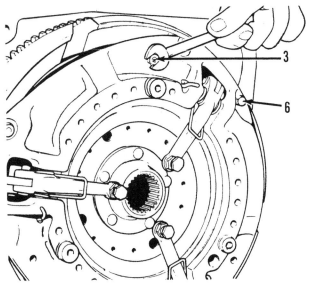

Fig. 108—Three 1/4 UNC screws (3) should be installed as shown to hold spring pressure when removing or installing dual clutch. Six screws (6) attach cover to flywheel.

screws to 27-34 N•m (20-25 ft.-lbs.) torque, then re-move the three compression screws (3).

Check clearance (C—Fig. 107) between heads of the three pto adjusting screws and plate. Clearance should be 2 mm (0.080 inch) and can be adjusted by turning screw after loosening lock nut. Clearance at all three locations should be exactly the same for even engagement. Use setting gauge (MF 314 or equivalent) to measure height of release fingers which should be 82.31 - 82.55 mm (3.25 - 3.28 inches). Adjust screws in fingers so that distance from back of screw to flywheel is correct and all are exactly the same. Refer to paragraph 93 and reconnect engine to transmission housing. Readjust clutch linkage as outlined in paragraph 92.

95. OVERHAUL. Refer to paragraph 94 and remove unit from tractor. Set clutch assembly in a press and compress assembly until the three screws (3—Fig. 108) are free and can be easily removed. Remove cotter pins (8—Fig. 109), pins (6) and springs (7), then disconnect levers (3) from links (20). Release pressure from press slowly and allow springs (14) to fully extend. Lift cover (2) and Belleville spring (16) from clutch, then lift pto pressure plate (17) and pto clutch disc (18) from clutch. Remove false flywheel (19), springs (14) and fiber washers (13).

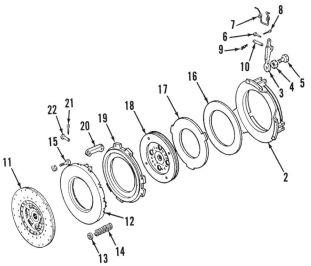

Fig. 109—Exploded view of dual clutch typical of models so equipped.

2. Clutch cover	13. Fiber washer
3. Release lever (3 used)	14. Engine clutch spring
4. Lock nut (3 used)	15. Pto adjusting screw
5. Screw (3 used)	and lock nut (3 used)
6. Pin (3 used)	16. Pto Belleville spring
7. Lever spring (3 used)	17. Pto pressure plate
8. Pin (3 used)	18. Pto friction disc
9. Pin (3 used)	19. False flywheel
10. Pivot pin (3 used)	for pto clutch
11. Engine clutch	20. Connecting links (3 used)
friction disc	21. Pin (3 used)
12. Engine clutch	22. Pin (3 used)
pressure plate	

Check all parts for wear, scoring or evidence of overheating. Friction surfaces of false flywheel (19), pressure plate (17) and pressure plate (12) should not be resurfaced. If damaged, new parts should be installed. Friction surface of engine flywheel can be surfaced, but no more than a total of 1.0 mm (0.040 inch) can be removed. The ledge on flywheel, to which the clutch is attached, should have exactly the same amount removed as the friction surface. Usually only 0.25 mm (0.010 inch) will need to be skimmed from surfaces to restore flywheel. Refer to Fig. 110 for flywheel dimensions (A and D).

Refer to Fig. 109 and the following:

Engine friction disc (11) –
   Diameter . . . . 305 mm (12 inches)
   Thickness . . . . 9.22 mm (0.363 inch)
     Wear limit . . . . 6.72 mm (0.265 inch)
Pto friction disc (18) –
   Diameter . . . . 254 mm (10 inches)
   Thickness . . . . 8.00 mm (0.315 inch)
     Wear limit . . . . 6 mm (0.236 inch)
Engine clutch pressure springs (14) –
  362, 365 & 375 Models
    Color . . . . Green
    Number used 9
    Clamp load . . . . 749 kg (1650 lbs.)
  383 & 390 Models
    Color . . . . Red
    Number used 12
    Clamp load . . . . 885 kg (1950 lbs.)
Pto Belleville spring (16) –
  362, 365 & 375 Models
    Color . . . . Dark Green
    Clamp load . . . . 749 kg (1650 lbs.)
  383 & 390 Models
    Color . . . . Orange
    Clamp load . . . . 885 kg (1950 lbs.)

Compress clutch in a press while assembling. Springs should be located as shown in Fig. 111. Refer to paragraph 94 for adjustment and to paragraph 93 for installation and reassembly. Refer to paragraph 92 to adjust linkage.

All Models With Split Torque Clutch

96. REMOVE AND REINSTALL. To remove the clutch assembly, first refer to paragraph 93 and separate engine from transmission housing. Install three ¼ UNC screws 2⅛ inches (54 mm) long into the three equi-spaced holes (3—Fig. 112) in clutch cover to hold. Loosen the six screws (6) retaining the clutch to flywheel, then remove screws and separate clutch from flywheel. **Be careful to avoid breathing dust which may contain asbestos. Do not use compressed air to blow dust from clutch and flywheel.**

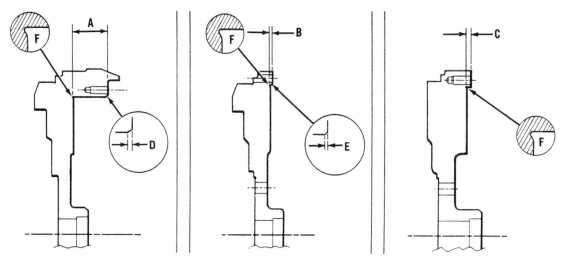

Fig. 110—The same amount of material machined from friction surface of engine flywheel should be removed from ledge to which clutch is attached to maintain distance (A, B or C). Angle of chamfer (D and E) should be 45°. Original contour (F) should be maintained.

A. 39.62-39.75 mm
(1.560-1.565 inches) B. 3.3 mm (0.130 inch) C. 3.3 mm (0.130 inch) D. 1.6 mm (0.063 inch)
E. 0.5 mm (0.020 inch)

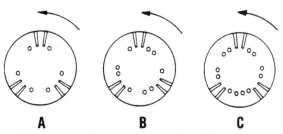

A   B   C

Fig. 111—Engine clutch springs should be located as shown, depending upon whether 6, 9 or 12 springs are used.

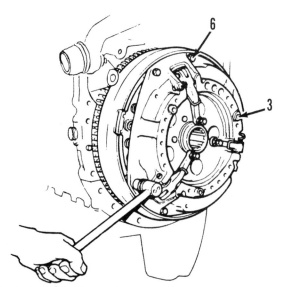

Fig. 112—Three 1/4 UNC screws (3) should be installed as shown to hold spring pressure when removing or installing split torque clutch. Six screws (6) attach cover to flywheel.

Position main clutch disc on flywheel and center using alignment tool (MF 159B or equivalent). Make sure three screws (3) are installed and tightened to compress springs, then locate clutch assembly and install the six retaining screws (6). Tighten retaining screws to 27-34 N•m (20-25 ft.-lbs.) torque for 305 mm (12 inches) diameter clutches, or 36-54 N•m (26-40 ft.-lbs.) for 330 mm (13 inches) diameter clutches, then remove the three compression screws (3).

On models with 305 mm (12 inches) diameter split torque clutch, use setting gauge (MF 314 or equivalent) to measure height of release fingers which should be 82.31-82.55 mm (3.25-3.28 inches). On 390 and 398 models with 330 mm (13 inches) diameter clutch, use setting gauge (MF 446 or equivalent) to measure height of release fingers which should be 118.92-120.22 mm (4.682-4.733 inches). On all models, adjust screws in fingers so that distance from back of screw to flywheel is correct and all are exactly the same. Refer to paragraph 93 and reconnect engine to transmission housing. Readjust clutch linkage as outlined in paragraph 92.

97. OVERHAUL. Refer to paragraph 96 and remove unit from tractor. Set clutch assembly in a press and compress assembly until the three screws (3—Fig. 112) are free and can be easily removed. Remove cotter pins from pins (6—Fig. 113 or Fig. 114), then remove pins and springs (7) Release pressure from press slowly and allow springs (14) to fully extend, then lift cover (2) from clutch. Remove springs (14) and washers (13). Hub (18) can be removed if necessary after removing nuts (17) and screws (19).

Check all parts for wear, scoring or evidence of overheating. Friction surface of pressure plate (12) should not be resurfaced. If damaged, new part

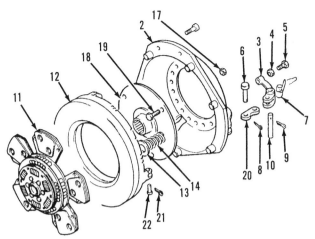

Fig. 113—Exploded view of split torque clutch typical of type used on some models. Refer also to Fig. 114.

2. Clutch cover
3. Release lever (3 used)
4. Lock nut (3 used)
5. Screw (3 used)
6. Pin (3 used)
7. Lever spring (3 used)
8. Pin (3 used)
9. Pin (3 used)
10. Pivot pin (3 used)
11. Engine clutch friction disc
12. Engine clutch pressure plate
13. Washer
14. Engine clutch spring
17. Nut (3 used)
18. Hub
19. Screw (3 used)
20. Connecting links (3 used)
21. Pin (3 used)
22. Pin (3 used)

should be installed. Friction surface of engine flywheel can be surfaced, but no more than a total of 1.0 mm (0.040 inch) can be removed. The ledge on flywheel, to which the clutch is attached, should have exactly the same amount removed as the friction surface. Usually only 0.25 mm (0.010 inch) will need to be skimmed from surfaces to restore flywheel. Refer to Fig. 110 for flywheel dimensions (B, C and E).

Refer to Fig. 113 or Fig. 114 and the following:

Engine Friction Disc (11)
Diameter
    All Models Except 398 . . . . . . 305 mm (12 inches)
    398 Models . . . . 330 mm (13 inches)
Thickness (All Models) . . . . . . 9.22 mm (0.363 inch)
    Wear limit . . . . 6.72 mm (0.265 inch)
Engine Clutch Pressure Springs (14) With Fiber (Not Cerametalic) Lining –
362 & 365 Models
    Color . . . . Green
    Number used 9

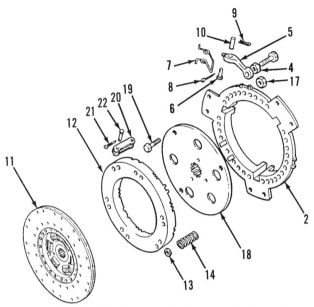

Fig. 114—Exploded view of split torque clutch typical of type used on some models. Refer to text for specifications of all models. Refer to Fig. 113 for legend.

    Clamp load . . . . 749 kg (1650 lbs.)
390 Models
    Color . . . . Red
    Number used 12
    Clamp load . . . . 885 kg (1950 lbs.)
398 Models
    Color . . . . Green
    Number used 12
    Clamp load . . . . 998 kg (2200 lbs.)
Engine Clutch Pressure Springs (14) With Five Pad Cerametalic Lining –
305 mm (12 inches) Diameter
    Color . . . . Red
    Number used 9
    Clamp load . . . . 664 kg (1464 lbs.)
330 mm (13 inches) Diameter
    Color . . . . Red
    Number used 12
    Clamp load . . . . 885 kg (1950 lbs.)

Compress clutch in a press while assembling. Springs should be located as shown in Fig. 111. Tighten screws (19—Fig. 113 or Fig. 114) to 34-41 N•m (25-30 ft.-lbs.) torque. Refer to paragraph 96 for adjustment, installation and reassembly. Refer to paragraph 92 to adjust linkage.

TRANSMISSION
(EIGHT-SPEED STANDARD)

365, 375, 383 and 390 Models

Models 365, 375, 383 and 390 are available with a transmission providing eight forward, non-synchronized speeds and two reverse speeds. The eight forward and two reverse speeds are possible by using a standard, sliding gear gearbox with four forward and one reverse speed compounded by a two speed planetary unit. The main gearbox of some models is equipped with overdrive in place of the standard 3rd gear. Refer to Fig. 115 for shift patterns of both standard sliding gear eight-speed transmission and overdrive models.

LUBRICATION

Models With Eight-Speed Standard Transmission

98. The transmission, rear differential and hydraulic system share common fluid which is contained in the transmission and center housing. Recommended lubricant is Massey-Ferguson Super 500 multi-use 10W-30 oil, Massey-Ferguson Permatran oil or equivalent transmission/hydraulic oil. Capacity is 47.4 L (12.5 gal.) if equipped with spacer or four-wheel-drive transfer box between transmission and rear axle center housing; 43.4 L (11.5 gal.) if not equipped with transfer box or spacer. Tractors are equipped with dipstick (Fig. 116 or Fig. 117) with marks indicating maximum and minimum oil levels. Level should be maintained at maximum level if tractor is operated on hilly terrain or if using implements which require large quantities of oil.

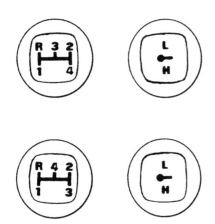

Fig. 115—Shift pattern for standard eight-speed transmission is shown in top view; shift pattern for eight-speed with overdrive in fourth speed of the main transmission is shown in lower view.

TRACTOR REAR SPLIT

Models With Eight-Speed Standard Transmission Without Cab

99. To separate tractors without cabs between the rear of the transmission housing and the front of the rear axle center housing, proceed as follows:

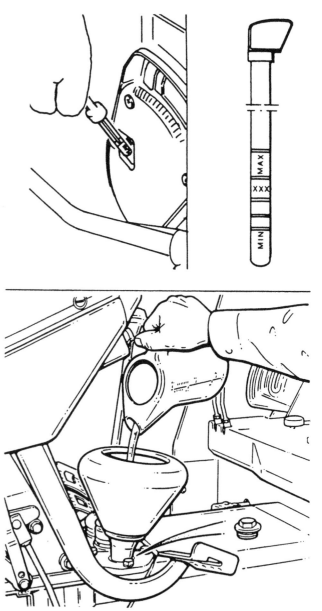

Fig. 116—Transmission oil level of tractors without cab is measured by dipstick located in the side cover as shown. Refer to Fig. 117 for models with cab.

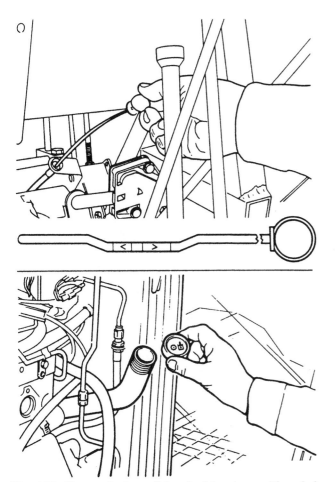

Fig. 117—Transmission oil level of tractors with cab is measured by dipstick that points toward rear as shown.

Set parking brake and block the rear wheels to prevent rolling. Disconnect battery ground cable and drain all transmission fluid. If equipped with four-wheel drive, refer to paragraph 9 and remove the front drive shaft. On all models, remove lower (gearbox) cover, both foot steps and battery boxes. Detach foot throttle, four-wheel-drive indicator switch and four-wheel-drive selector lever. Disconnect hydraulic oil supply pipe located on right side of tractor and, if so equipped, detach hydraulic pipes from left side. Disconnect brake lines and drain system. Disconnect wiring harness at the connectors below console on right side and disconnect wires from safety start switch. Remove the hydraulic suction filter assembly. Place hardwood wedges between the front axle and axle support casting on both sides to prevent tipping around axle pivot when tractor is separated. Support rear of tractor under axle center housing and front of tractor under transmission housing. Remove screws attaching rear axle center housing or four-wheel-drive transfer case to the transmission housing, then separate the tractor halves. Either the front can be moved forward or the rear can be moved to the rear, but it is important to support both halves safely and securely.

If the split pin is removed from the shear tube, install pin in center of the five holes. Install shear tube on pinion. Install at least two dowel studs to facilitate alignment, position new gasket on dowel studs and position rear drive shaft into high/low gear coupler. Move the high/low shifter to "low" and shift main transmission to 3rd. Move tractor together, turning engine to align splines of shear tube. Make sure that flanges are tight against each other before installing retaining screws. Install and tighten one screw on each side, then check end float of the rear drive shaft. Push the shear tube toward rear as shown at inset in Fig. 118, which will compress spring, then measure clearance between the shear tube and locking collar. Correct clearance is 0.4-2.5 mm (0.015-0.100 inch) and can be changed by separating tractor and moving the split ring to another hole in the shear tube. Tighten all screws retaining the four-wheel-drive transfer case, spacer housing or rear axle center housing to the transmission housing to 112 N•m (83 ft.-lbs.) torque, beginning at top center and progressing in a clockwise direction (as viewed from tractor rear) two times around the mating flange. If equipped with four-wheel drive, coat threads of drive shaft retaining screws with "Loctite 270," then attach drive shaft flanges and tighten screws to 55-75 N•m (40-55 ft.-lbs.) torque. Remainder of assembly is reverse of disassembly.

Models With Eight-Speed Standard Transmission With Cab

100. To separate tractors with cabs between the rear of the transmission housing and front of the rear axle center housing, proceed as follows:

Set parking brake, block the rear wheels to prevent rolling and drain all transmission fluid. Disconnect battery ground cable, then disconnect wiring harness at the connectors. If equipped with four-wheel drive, refer to paragraph 9 and remove the front drive shaft. On all models, disconnect throttle stop cable and throttle cable from fuel injection pump and move cables out of the way so that cables will not catch when separating tractor. Disconnect and cap all hydraulic pipes running from front to rear, which would prevent separation of tractor halves. Mark (identify), disconnect and cap steering hoses, heater hoses and air conditioning lines (if so equipped), which would interfere. Place hardwood wedges between the front axle and axle support casting on both sides to prevent tipping around axle pivot when tractor is separated. Remove front floor mat and floor plates, disconnect clutch linkage and detach gearbox oil filler pipe. Unbolt and remove the gear shift cover complete with shift levers, then unbolt and remove the transmission cover. Disconnect wires at safety start switch and four-wheel-drive indicator switch. Remove the oil

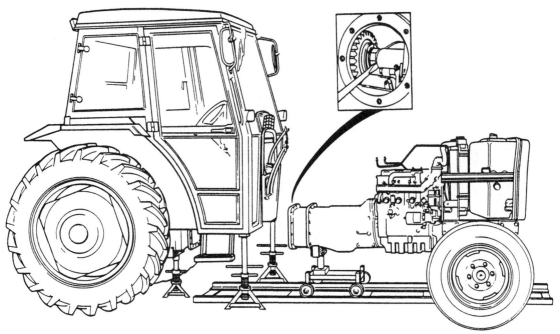

Fig. 118—View of tractor with cab separated between the four-wheel-drive transfer case and the rear axle center housing. Special splitting stand and track is shown. Support and splitting procedures are similar for models without cab and without four-wheel drive.

feed pipe and filter housing from the right side of transmission case. Disconnect four-wheel-drive selector lever, then check to make sure that all wiring, cables, linkages and hoses are free to permit the tractor to be separated. Support rear of tractor under axle center housing and at front corners of cab. Support front of tractor under transmission housing in such a way that front of tractor, engine, transmission and transfer case can be rolled forward (Fig. 118). Unbolt and remove both cab support brackets from front of cab. Remove screws attaching transfer case to the rear axle center housing, then move front of tractor forward. It is important to support both halves safely and securely to prevent damage and injury.

If the split pin is removed from the shear tube, install pin in center of the five holes. Install shear tube on pinion. Install at least two dowel studs to facilitate alignment, position new gasket on dowel studs and position rear drive shaft into high/low gear coupler. Move the high/low shifter to Low and shift main transmission to 3rd. Move front half of tractor toward rear turning engine to align splines of shear tube. Make sure that flanges are tight against each other before installing retaining screws. Install and tighten one screw on each side, then check end float of the rear drive shaft. Push the shear tube toward rear as shown at inset in Fig. 118, which will compress spring, then measure clearance between the shear tube and locking collar. Correct clearance is 0.4-2.5 mm (0.015-0.100 inch) and can be changed by separating tractor and moving the split ring to an-

other hole in the shear tube. Tighten all screws retaining the transmission housing to spacer housing, transfer case or rear axle center housing to 112 N·m (83 ft.-lbs.) torque, beginning at top center and progressing in a clockwise direction (as viewed from tractor rear) two times around the mating flange. On four-wheel-drive models, coat threads of drive shaft retaining screws with "Loctite 270," then attach drive shaft flanges and tighten screws to 55-75 N·m (40-55 ft.-lbs.) torque. Remainder of assembly is reverse of disassembly.

TRANSMISSION REMOVAL

All Models With Eight-Speed Standard Transmission

101. Refer to paragraph 99 or paragraph 100 to separate transmission housing from rear axle center housing. If not equipped with cab, unbolt and support or remove the steering and instrument console. On all models, support the engine and transmission housing separately. Remove screws and stud nuts, then separate transmission from engine.

Reassemble transmission to engine by reversing the splitting procedure and observe the following. Install a M12 alignment stud approximately 100 mm (4 inches) long in each side of the transmission housing to assist in aligning engine to transmission. Turn flywheel to align clutch plate splines with transmission and pto input shaft splines. Install retaining screws and nuts after engine and transmission housing flanges are completely together. Tighten retaining screws and nuts to 115 N·m (85 ft.-lbs.) torque.

Refer to paragraph 99 or 100 and reconnect transmission to rear axle center housing. Tighten all screws retaining the transmission housing to spacer housing, transfer case or rear axle center housing to 112 N·m (83 ft.-lbs.) torque, beginning at top center and progressing in a clockwise direction (as viewed from tractor rear) two times around the mating flange. On four-wheel-drive models, coat threads of drive shaft retaining screws with "Loctite 270," then attach drive shaft flanges and tighten screws to 55-75 N·m (40-55 ft.-lbs.) torque. On all models, check and adjust clutch linkage as outlined in paragraph 92.

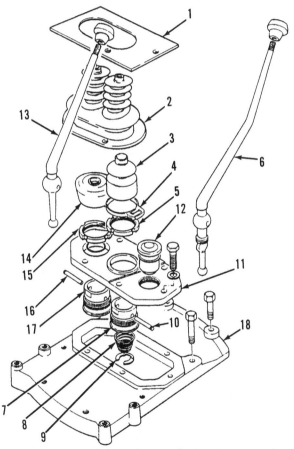

Fig. 119—Exploded view of transmission top cover, tower cover and shift levers typical of models with standard eight-speed transmission.

1. Panel	10. Pin
2. Boot	11. Tower cover
3. Cover	12. Fill plug
4. Retainer clip	13. High/low shift lever
5. Retaining nut	14. Cover
6. Main shift lever	15. Retaining nut
7. Sleeve	16. Pin
8. Spring	17. Sleeve
9. Snap ring	18. Top cover

TOP COVER AND SHIFT LEVERS

Models With Eight-Speed Standard Transmission

102. The shift levers and the small shift tower can be removed after first removing floor mat and access panels. Installation of the shift levers and tower is easier if transmission is shifted into neutral before removing. Shift tower (11—Fig. 119) is attached to the top cover (18) by five screws.

The main shift lever (6) can be removed after removing the tower assembly, cover (3), snap ring (9), spring (8) and pin (10). The high/low shift lever can be removed after removing cover (14) and pin (16).

The transmission top cover (18) can be unbolted and lifted from transmission after first removing the shift levers and tower assembly. On models with cab, it is necessary to split rear of transmission from rear axle center housing as outlined in paragraph 100 or remove the cab assembly. Models without cab must have instrument console and other interfering parts removed before transmission top cover. The reason for removing top cover will determine what additional procedures are necessary.

"Loctite 515 Instant Gasket" or equivalent should be used to seal tower to top cover. Tighten retaining screws to 50-70 N·m (37-52 ft.-lbs.) torque.

SHIFTER RAILS AND FORKS

Models With Eight-Speed Standard Transmission

103. To remove the shifter rails and forks, first refer to paragraph 99 or paragraph 100 and separate tractor between rear of transmission and front of axle center housing. Refer to paragraph 102 and remove the shift lever and tower assembly and transmission top cover. Models without cab must have instrument and steering console removed before top cover can be unbolted and removed.

On all models, remove the neutral safety start switch (N—Fig. 121). Remove safety wire and set screws (5—Fig. 120) and lift detent springs and plungers (4) from bores in housing. Unbolt and remove interlock parts (12, 13, 14 and 15—Fig. 121). Slide rail out of housing toward rear and remove gate and shift fork through top. Twist rails if necessary while withdrawing.

When assembling, lubricate rails and make sure that rails do not have burrs, especially around the set screw locations, that would prevent easy and smooth installation. Each of the four shift rails is different from the other rails. Tighten screws retaining the interlock retainer (15—Fig. 120) to 40-47 N·m (30-35 ft.-lbs.) torque and set screws (5) to 34-52 N·m (25-38 ft.-lbs.) torque. Install safety wire through set screws (5) after tightening to prevent loosening. Shift all gears, rails and gates to neutral after checking to make sure that each gear can be correctly engaged. Install detent plungers and springs (4). Remainder of assembly is the reverse of disassembly procedure.

PLANETARY UNIT

Models With Eight-Speed Standard Transmission

104. To remove the planetary unit, first refer to paragraph 99 or paragraph 100 and separate tractor between rear of transmission and front of axle center housing. Remove set screw (5—Fig. 120) from fork (3), then remove the fork and coupler from rear of plane-

tary. Remove the four retaining cap screws and withdraw rear cover (13—Fig. 122 or Fig. 123), rear thrust ring (6) and planet carrier (7). Pry planetary ring gear (3) with dowels from rear of housing. Remove planetary front cover (2) and shim (1).

To disassemble planet carrier (7), first remove snap ring (5—Fig. 123), if used. Press planet pins (8—Fig. 122 or Fig. 123) forward out of carrier. Remove planet gears (10) with needle rollers (12) and thrust washers (9). Splined sleeve (15) is retained in hub of carrier (7) by retaining ring (16).

When reassembling, be sure to account for all of the needle rollers (12). Each pinion contains two paths of rollers separated by a washer (11). Each roller path contains 27 rollers (12—Fig. 122) on 365 models. Each roller path contains 16 rollers (12—Fig. 123) on 375, 383 and 390 models. Use petroleum jelly (not heavy grease) to hold rollers in place while pressing

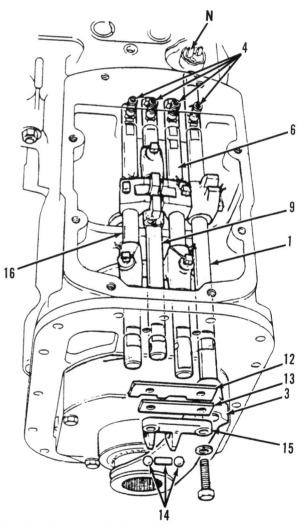

Fig. 120—Exploded view of shift rails and forks for standard eight-speed transmission. Because of gear differences, shift rail (9) is for fourth speed and rail (16) is for second and third speeds, of models with overdrive.

1. Shift rail (High/Low)	10. Shift fork
2. Shift gate	11. Shift gate
3. Shift fork	12. Plate
4. Detent springs and plungers	13. Gasket
5. Retaining screws	14. Interlock plunger & balls
6. Shift rail (Rev./1st)	15. Retainer
7. Shift gate	16. Shift rail (2nd/4th)
8. Shift fork	17. Shift gate
9. Shift rail (3rd)	18. Shift fork

Fig. 121—Drawing of shift rails, installed. Notice position of notches in plate (12) and location of neutral start switch (N). Plunger of interlock (14) is located in rail (9) and balls are located in retainer (15). Refer to Fig. 120 for legend.

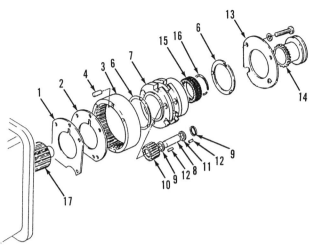

Fig. 122—Exploded view of high/low range planetary unit used on 365 model. Refer to Fig. 123 for other models with eight-speed standard transmission.

1. Front shim	10. Pinion
2. Front plate	11. Spacer washer
3. Ring gear	12. Needle rollers (27/path)
4. Dowel	13. Rear plate
6. Thrust washer	14. Shift coupler
7. Planet carrier	15. Splined sleeve
8. Pinion shaft	16. Retaining ring
9. Side washer	17. Transmission mainshaft

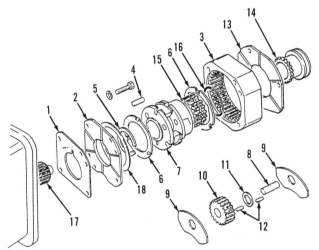

Fig. 123—Exploded view of high/low range planetary unit used on 375, 383 and 390 models. Refer to Fig. 184 for planetary unit used on other models with eight-speed standard transmission.

1. Front shim	10. Pinion
2. Front plate	11. Spacer washer
3. Ring gear	12. Needle rollers (16/path)
4. Dowel	13. Rear plate
5. Snap ring	14. Shift coupler
6. Thrust washer	15. Splined sleeve
7. Planet carrier	16. Retaining ring
8. Pinion shaft	17. Transmission mainshaft
9. Side washer	18. Belleville washer

planet pins into carrier. Gap in snap ring (5—Fig. 123) should be located between planet pins.

Concave side of Belleville washer (18—Fig. 123) should be toward rear. Be sure that front plate (2—Fig. 122 or Fig. 123) and rear plate (13) are positioned with grooved side toward planet carrier. Slots in front plate and shim (1) must be aligned and toward top when installed on transmission. Use petroleum jelly to hold thrust rings (6) in place. Make certain that tangs of thrust rings engage notches in pinion carrier and that brass side faces away from carrier. On 365 models, cutaway section of rear plate (13—Fig. 122) should be positioned at the lower, **right** corner of planetary as shown. A lockwasher is used only on mounting screw at corner of cutaway. On 375, 383 and 390 models the slanted section of rear plate (13—Fig. 123) should be positioned at lower, **left** corner of planetary. On all models, tighten mounting screws to 40-47 N•m (30-35 ft.-lbs.) torque. Complete assembly procedure by reversing disassembly.

PTO INPUT SHAFT AND RETAINER HOUSING

Models With Eight-Speed Standard Transmission

105. To remove the pto input shaft (8—Fig. 124) and retainer (3), first split tractor between engine and transmission as outlined in paragraph 93. Remove clutch release bearing (16), retainer (15), fork (13) and release shaft (12). Remove brake cross shaft, if so equipped. Remove screws and pull pto input shaft (8) and retainer (3) as a unit from transmission housing.

To disassemble, remove snap ring (6) from rear of retainer, then push input shaft and bearing (5) out of housing toward rear. Bearing can be removed from input shaft after removing snap ring (4). Remove oil seals (1 and 9) and needle bearing (2) only if new parts are to be installed.

Lubricate all seals and bearings with petroleum jelly before assembling. Use special tool (MF 315A or equivalent) to install new needle bearing (2). Install seal (9) with lip toward rear, using special tool (MF 331 or equivalent). Install ball bearing (5) on shaft with shield toward rear, then install snap ring (4). Use special tool (MF255B or equivalent) to install seal (1) with lip toward rear. Use special sleeve (KMF 1004 or equivalent) to protect seal (1), slide input shaft into the retainer housing, then install snap ring (6). Install "O" ring (7) and install housing and shaft assembly over transmission input shaft (11) into housing bore. Coat threads of retaining screws with "Hylomar" or equivalent sealer and tighten to 54-61 N•m (40-45 ft.-lbs.) torque. Remainder of assembly is reverse of disassembly.

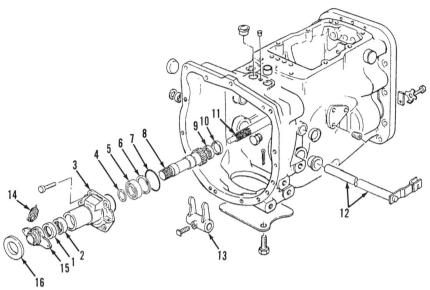

Fig. 124—Pto input shaft (8) and retainer housing (3) can be removed as a unit from front of transmission housing. Transmission input shaft (11) is removed rearward from transmission housing.

1. Seal
2. Needle bearing
3. Retainer housing
4. Snap ring

5. Ball bearing
6. Snap ring
7. "O" ring
8. Pto input shaft

9. Oil seal
10. Thrust washer
11. Transmission input shaft
12. Clutch lever and shaft

13. Clutch release lever
14. Return spring
15. Release housing
16. Release bearing

PTO LOWER SHAFT FRONT BEARING AND RETAINER

Models With Eight-Speed Standard Transmission

106. The pto lower shaft front bearing (41—Fig. 125) can be removed and installed after separating tractor between the rear of the engine and front of

transmission housing. Removal of the lower driven gear (45) and shaft (46) require the additional steps of separating at rear of transmission housing and removal of the transmission top cover and shift rails. The following procedure describes only service to the front bearing and bearing retainer (42) after splitting as outlined in paragraph 93.

Remove clutch release bearing (16—Fig. 124), retainer (15), fork (13) and release shaft (12). Unbolt and remove cover (36—Fig. 125), then remove snap

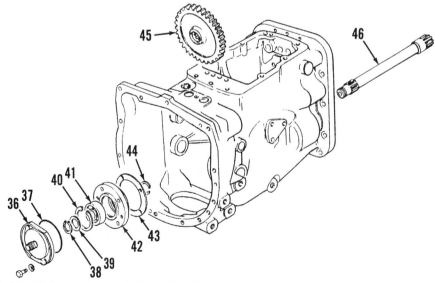

Fig. 125—Exploded view of the pto lower shaft and bearings.

36. Cover
37. "O" ring
38. Snap ring

39. Washer
40. Snap ring
41. Ball bearing

42. Retainer
43. Gasket

44. Snap ring
45. Pto lower gear
46. Pto lower shaft

ring (38) and washer (39). Thread a ⅜ inch UNC cap screw, 75 mm (3 inches) long into each of the two threaded holes of retainer (42). Tighten the two jackscrews to push retainer (42) and bearing (41) forward off shaft (46). Bearing can be pressed from retainer after removing snap ring (40). Remove "O" ring (37) and gasket (43), then clean all parts, thoroughly.

Coat gasket (43) and threads of retaining screws lightly with "Hylomar" or equivalent sealer before installing. End of shaft (46) is threaded to accept a puller screw to pull bearing (41) onto shaft until washer (39) and snap ring (38) can be installed. Align holes in retainer (42) with holes in transmission housing, install "O" ring (37) and cover (36). Tighten retaining screws to 54-61 N·m (40-45 ft.-lbs.) torque. Remainder of assembly is reverse of disassembly procedure.

TRANSMISSION INPUT SHAFT

Models With Eight-Speed Standard Transmission

107. To remove the transmission input shaft (11—Fig. 124), first remove the transmission as outlined in paragraph 101. Remove the top cover and shift levers as outlined in paragraph 102, planetary unit as outlined in paragraph 104 and the shift rails and forks as outlined in paragraph 103. Refer to paragraph 106 and remove the pto lower shaft front bearing and retainer. Remove snap ring (44—Fig. 125), pull shaft (46) to rear and allow gear (45) to fall to bottom of housing. Move input shaft (11—Fig. 126) and gear (13) forward, then withdraw gear (13) out top. Transmission input shaft (11) can be removed from same top opening after gear (13) is out.

Fig. 126—Exploded view of mainshaft, countershaft and gears. All transmission input shafts (11) do not have a separate gear (11G) located on splines. Refer to Fig. 124 for pto input shaft and Fig. 125 for pto lower shaft. Some transmissions have different ratio gears at (13 and 33), which makes this gear selection faster than gears (21 and 28).

11. Transmission input shaft
11G. Separate gear
12. Spacer
13. Gear (3rd)
14. Snap ring
15. Ball bearing
16. Snap ring
17. Spacer
18. Roller bearing
19. Main shaft
20. Gear (1st & Reverse)
21. Gear (2nd & 4th)
22. Ball bearing
23. Snap rings
25. Snap ring
26. Ball bearing
27. Gear (2nd)
28. Gear (4th)
29. Snap ring
30. Roller bearing
31. Countershaft & 1st gear
32. Ball bearing
33. Gear (3rd)
34. Constant mesh gear
35. Snap ring
50. Thrust washers
51. Washers
52. Center spacer
53. Needle rollers
54. Gear (Reverse idler)
56. Spacer
57. Idler shaft

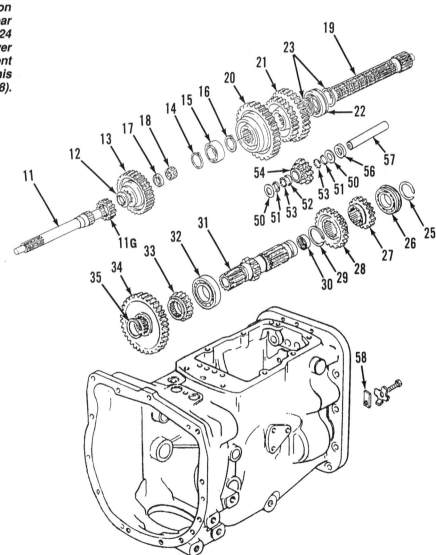

Reinstall input shaft by reversing removal procedures.

MAINSHAFT (OUTPUT SHAFT) AND GEARS

Models With Eight-Speed Standard Transmission

108. To remove the mainshaft (19—Fig. 127), first remove the transmission as outlined in paragraph 101. Remove the top cover and shift levers as outlined in paragraph 102, planetary unit as outlined in paragraph 104 and the shift rails and forks as outlined in paragraph 103. Refer to paragraph 105 and remove the pto input shaft and retainer, remove the pto lower shaft front bearing and retainer as outlined in paragraph 106, then remove the transmission input shaft as outlined in paragraph 107.

Remove spacer (12—Fig. 126), drive the mainshaft (19) to rear and remove gear (13). Remove snap ring (14) and continue to drive mainshaft toward rear out of bearing (15). Remove snap ring (16) and withdraw mainshaft toward rear while lifting gears (20 and 21) out top opening. Rear bearing (22) can be removed if necessary after removing snap rings (23).

Reassemble by reversing removal procedure. Refer to Fig. 127 for cross section of assembled transmission. Coat gasket (43—Fig. 125) lightly with "Hylomar" or equivalent sealer before installing. Coat threads of screws attaching cover (36) and screws attaching retainer housing (3—Fig. 124) lightly with "Hylomar" or equivalent sealer before tightening to 54-61 N·m (40-45 ft.-lbs.) torque.

COUNTERSHAFT (LAYSHAFT) AND GEARS

Models With Eight-Speed Standard Transmission

109. The mainshaft must be removed as outlined in paragraph 108 before countershaft. Remove snap ring (25—Fig. 126) from rear, bump shaft forward enough to remove snap ring (29) from groove and move snap ring forward to unsplined part of countershaft. Remove snap ring (35), then remove gears (33 and 34) while driving countershaft toward rear. After gears (33 and 34) are lifted from housing, drive countershaft toward front while removing gears (27 and 28).

Reassemble by reversing removal procedure. Snap ring (29) should be around shaft in the unsplined area and front bearing (32) should be in place on shaft before installing shaft in housing. Refer to Fig. 127 for cross section of assembled transmission.

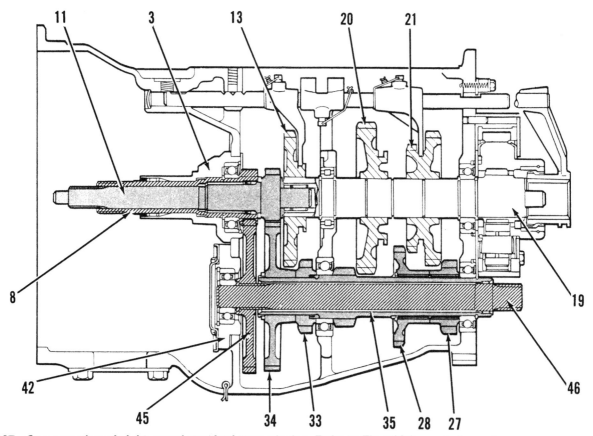

Fig. 127—Cross section of eight-speed standard transmission. Refer to Fig. 126 for parts legend.

REVERSE IDLER GEAR AND SHAFT

Models With Eight-Speed Standard Transmission

110. A dummy shaft should be fabricated to hold parts in place while both removing and installing the reverse idler shaft. Dummy shaft should be 25 mm (1 inch) diameter and 55 mm (2 3/16 inches) long.

To remove the reverse idler, first remove the mainshaft and countershaft assemblies as outlined in paragraphs 108 and 109. Release locking washer, then remove locking screw and clip (58—Fig. 126). Slide dummy shaft in from front, pushing reverse idler shaft (57) out toward rear. Dummy shaft should be used to hold needle rollers and washers in place while both removing and installing the reverse idler. Be sure to remove all 56 of the needle rollers.

Assemble needle rollers, washers and spacer in reverse idler using petroleum jelly and the dummy shaft. Assemble center spacer (52), one path of 28 needle rollers (53), one end washer (51) and one thrust washer (50), around dummy shaft, in one end of reverse idler gear. Assemble the second path of 28 needle rollers, the other end washer (51) and the remaining thrust washer (50). Center the dummy shaft so that it holds both thrust washers in place, then position assembly in housing and install idler shaft (57). Install clip (58), lock plate and retaining screw. Tighten retaining screw and lock by bending tab of lock plate against flat of screw. Complete reassembly by reversing disassembly procedure.

TRANSMISSION (EIGHT-SPEED SYNCHROMESH)

362, 365, 375 and 390 Models

Models 362, 365, 375 and 390 are available with a transmission providing eight forward speeds and two reverse speeds. The eight forward and two reverse speeds are possible by using a main gearbox with four forward and one reverse speed compounded by a two speed planetary unit. The main gearbox is equipped with synchronized shifting in top two speeds. Refer to Fig. 128 for shift pattern of eight-speed synchronized transmission models.

LUBRICATION

Models With Eight-Speed Synchromesh Transmission

111. The transmission, rear differential and hydraulic system share common fluid which is contained in the transmission and center housing. Recommended lubricant is Massey-Ferguson Super 500 multi-use 10W-30 oil, Massey-Ferguson Permatran oil or equivalent transmissio/hydraulic oil. Capacity is 47.4 L (12.5 gal.) if equipped with spacer or four-wheel-drive transfer box between transmission and rear axle center housing; 43.4 L (11.5 gal.) if not equipped with transfer box or spacer. Tractors are equipped with dipstick (Fig. 129 or Fig. 130) with marks indicating maximum and minimum oil levels. Level should be maintained at maximum level if tractor is operated on hilly terrain or if using implements which require large quantities of oil.

TRACTOR REAR SPLIT

Models With Eight-Speed Synchromesh Transmission

112. To separate the rear of the transmission from the rear axle center housing (Fig. 131), refer to paragraph 99 if tractor does not have cab or to paragraph 100 if equipped with cab.

TRANSMISSION REMOVAL

Models With Eight-Speed Synchromesh Transmission

113. To remove the transmission, first separate the rear of the transmission from rear axle center housing as outlined in paragraph 99 if tractor does not have cab or to paragraph 100 if equipped with cab. If not equipped with cab, unbolt and support or remove the steering and instrument console. On all models, support the engine and transmission housing separately. Remove screws and stud nuts, then separate transmission from engine.

Fig. 128—Shift pattern for eight-speed synchronized transmission. Only third and fourth speeds of the main transmission are synchronized.

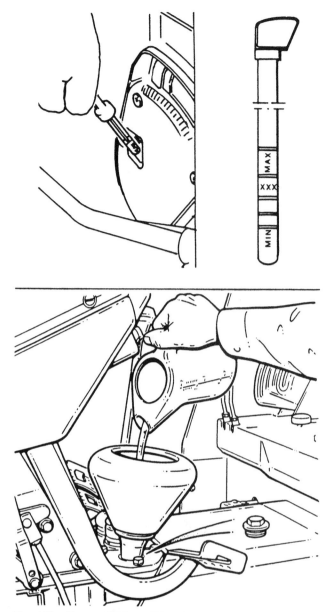

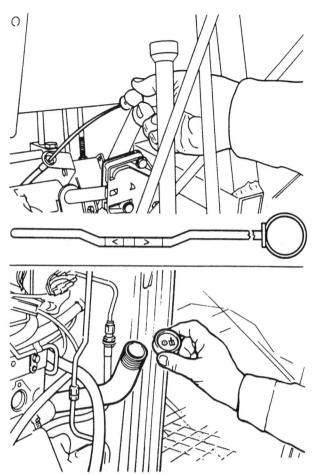

Fig. 130—Transmission oil level of tractors with cab is measured by dipstick that points toward rear as shown.

Fig. 129—Transmission oil level of tractors without cab is measured by dipstick located in the side cover as shown. Refer to Fig. 130 for models with cab.

Reassemble transmission to engine by reversing the splitting procedure and observe the following. Install a M12 alignment stud approximately 100 mm (4 inches) long in each side of the transmission housing to assist in aligning engine to transmission. Turn flywheel to align clutch plate splines with transmission and pto input shaft splines. Install retaining screws and nuts after engine and transmission housing flanges are completely together. Tighten retaining screws and nuts to 115 N·m (85 ft.-lbs.) torque.

Refer to paragraph 99 or 100 and reconnect transmission to rear axle center housing. Tighten all screws retaining the transmission housing to spacer housing, transfer case or rear axle center housing to 112 N·m (83 ft.-lbs.) torque, beginning at top center

and progressing in a clockwise direction (as viewed from tractor rear) two times around the mating flange. On four-wheel-drive models, coat threads of drive shaft retaining screws with "Loctite 270," then attach drive shaft flanges and tighten screws to 55-75 N·m (40-55 ft.-lbs.) torque. On all models, check and adjust clutch linkage as outlined in paragraph 92.

TOP COVER

Models With Eight-Speed Synchromesh Transmission

114. The shift levers and the small shift tower can be removed after first removing floor mat and access panels. Installation of the shift levers and tower is easier if transmission is shifted into neutral before removing. Shift tower (11—Fig. 132) is attached to the top cover (18) by five screws.

The main shift lever (6) can be removed after removing the tower assembly, cover (3), snap ring (9), spring (8) and pin (10). The high/low shift lever can be removed after removing cover (14) and pin (16).

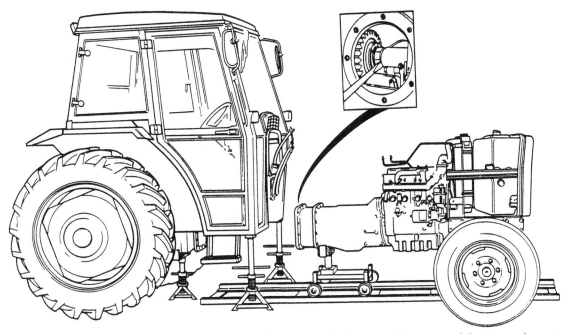

Fig. 131—View of tractor with cab separated between the four-wheel-drive transfer case and the rear axle center housing. Special splitting stand and track is shown. Support and splitting procedures are similar for models without cab and without four-wheel drive.

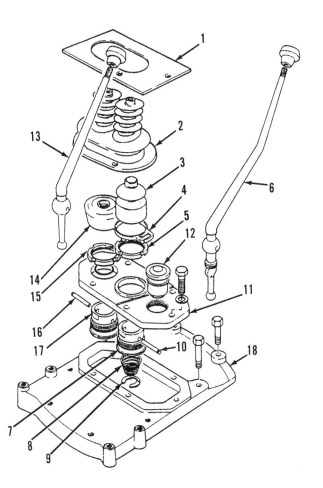

Fig. 132—Exploded view of transmission top cover, tower cover and shift levers typical of models with eight-speed transmission.

1. Panel
2. Boot
3. Cover
4. Retainer clip
5. Retaining nut
6. Main shift lever
7. Sleeve
8. Spring
9. Snap ring
10. Pin
11. Tower cover
12. Fill plug
13. High/low shift lever
14. Cover
15. Retaining nut
16. Pin
17. Sleeve
18. Top cover

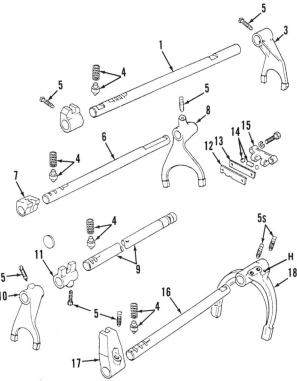

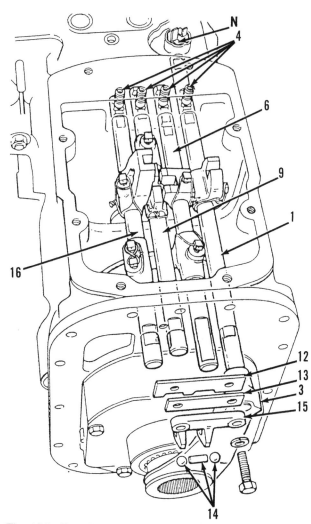

Fig. 133—Exploded view of shift rails and forks for eight-speed transmission with synchronized third and fourth speeds in the main transmission.

1. Shift rail (High/Low)
2. Shift gate
3. Shift fork
4. Detent springs and plungers
5. Retaining screws
6. Shift rail (Rev./1st)
7. Shift gate
8. Shift fork
9. Shift rail (2nd)
10. Shift fork
11. Shift gate
12. Plate
13. Gasket
14. Interlock plunger & balls
15. Retainer
16. Shift rail (3rd/4th)
17. Shift gate
18. Shift fork

Fig. 134—Drawing of shift rails, installed. Notice position of notches in plate (12) and location of neutral start switch (N). Plunger of interlock (14) is located in rail (9) and balls are located in retainer (15). Refer to Fig. 133 for legend.

The transmission top cover (18) can be unbolted and lifted from transmission after first removing the shift levers and tower assembly. On models with cab, it is necessary to split rear of transmission from rear axle center housing as outlined in paragraph 100 or remove the cab assembly. Models without cab must have instrument console and other interfering parts removed before transmission top cover. The reason for removing top cover will determine what additional procedures are necessary.

"Loctite 515 Instant Gasket" or equivalent should be used to seal tower to top cover. Tighten retaining screws to 50-70 N·m (37-52 ft.-lbs.) torque.

SHIFTER RAILS AND FORKS

Models With Eight-Speed Synchromesh Transmission

115. To remove the shifter rails and forks, first refer to paragraph 99 or paragraph 100 and separate trac-

tor between rear of transmission and front of axle center housing. Refer to paragraph 114 and remove the shift lever and tower assembly and transmission top cover. Models without cab must have instrument and steering console removed before top cover can be unbolted and removed.

On all models, remove the neutral safety start switch (N—Fig. 134). Remove safety wire and set screws (5—Fig. 133) and lift detent springs and plungers (4) from bores in housing. Unbolt and remove interlock parts (12, 13, 14 and 15—Fig. 134). Slide rail out of housing toward rear and remove gate and shift fork through top. Twist rails if necessary while withdrawing.

When assembling, lubricate rails and make sure that rails do not have burrs, especially around the set screw locations, that would prevent easy and smooth installation. Each of the four shift rails is different from the other rails. Tighten screws retaining the interlock retainer (15—Fig. 133) to 40-47 N·m (30-35 ft.-lbs.) torque and the seven set screws (5) to 34-52 N·m (25-38 ft.-lbs.) torque. Install safety wire through set screws (5) after tightening to prevent loosening. Adjust position of shift fork (18) as follows.

Use special tools (MF 414/2 and MF 414/3 or equivalent) to hold shift rail (9—Fig. 133 and Fig. 134) in neutral detent position. Insert centralizing tool (MF 414/1 or equivalent) into hole (H—Fig. 133) and into hole of sliding synchronizing coupling. Tighten each of the two set screws (5S), while turning centralizing tool, to hold the fork in this center position. Screws (5S) should also be tightened to 34-52 N·m (25-38 ft.-lbs.) torque and locked in position with safety wire. Shift all gears, rails and gates to neutral after checking to make sure that shift gates all align. Shift into each gear to make sure that each can be correctly engaged. Install detent plungers and springs (4—Fig. 134). Remainder of assembly is the reverse of disassembly procedure.

PLANETARY UNIT

Models With Eight-Speed Synchromesh Transmission

116. To remove the planetary unit, first refer to paragraph 99 or paragraph 100 and separate tractor between rear of transmission and front of axle center housing. Remove set screw (5—Fig. 133) from fork (3),

then remove the fork and coupler from rear of planetary. Remove the four retaining cap screws and withdraw rear cover (13—Fig. 135 or Fig. 136), rear thrust ring (6) and planet carrier (7). Pry planetary ring gear (3) with dowels from rear of housing. Remove planetary front cover (2) and shim (1).

To disassemble planet carrier (7), first remove snap ring (5—Fig. 136), if used. Press planet pins (8—Fig. 135 or Fig. 136) forward out of carrier. Remove planet gears (10) with needle rollers (12) and thrust washers (9). Splined sleeve (15) is retained in hub of carrier (7) by retaining ring (16).

When reassembling, be sure to account for all of the needle rollers (12). Each pinion contains two paths of rollers separated by a washer (11). Each roller path contains 27 rollers (12—Fig. 135) on 362 and 365 models. Each roller path contains 16 rollers (12—Fig. 136) on 375 and 390 models. Use petroleum jelly (not heavy grease) to hold rollers in place while pressing planet pins into carrier. Gap in snap ring (5) should be located between planet pins.

Concave side of Belleville washer (18—Fig. 136) should be toward rear. Be sure that front plate (2—Fig. 135 or Fig. 136) and rear plate (13) are positioned with grooved side toward planet carrier. Slots in front plate and shim (1) must be aligned and toward top when installed on transmission. Use petroleum jelly to hold thrust rings (6) in place. Make certain that tangs of thrust rings engage notches in pinion carrier and that brass side faces away from carrier. On 362

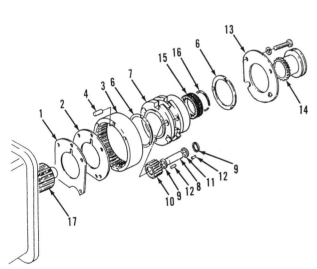

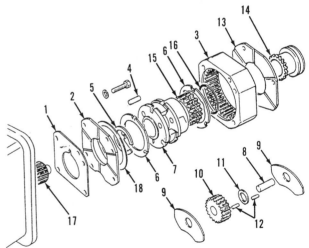

Fig. 135—Exploded view of high/low range planetary unit used on 362 and 365 models. Refer to Fig. 136 for other models with eight-speed synchronized transmission.

1. Front shim	10. Pinion
2. Front plate	11. Spacer washer
3. Ring gear	12. Needle rollers (27/path)
4. Dowel	13. Rear plate
6. Thrust washer	14. Shift coupler
7. Planet carrier	15. Splined sleeve
8. Pinion shaft	16. Retaining ring
9. Side washer	17. Transmission mainshaft

Fig. 136—Exploded view of high/low range planetary unit used on 375 and 390 models. Refer to Fig. 135 for planetary unit used on other models with eight-speed synchronized transmission.

1. Front shim	10. Pinion
2. Front plate	11. Spacer washer
3. Ring gear	12. Needle rollers (16/path)
4. Dowel	13. Rear plate
5. Snap ring	14. Shift coupler
6. Thrust washer	15. Splined sleeve
7. Planet carrier	16. Retaining ring
8. Pinion shaft	17. Transmission mainshaft
9. Side washer	18. Belleville washer

and 365 models, cutaway section of rear plate (13—Fig. 135) should be positioned at the lower, **right** corner of planetary as shown. A lockwasher is used only on mounting screw at corner of cutaway. On 375 and 390 models, the slanted section of rear plate (13—Fig. 136) should be positioned at lower, **left** corner of planetary. On all models, tighten mounting screws to 40-47 N•m (30-35 ft.-lbs.) torque. Complete assembly procedure by reversing disassembly.

PTO INPUT SHAFT AND RETAINER HOUSING

Models With Eight-Speed Synchromesh Transmission

117. To remove the pto input shaft (8—Fig. 137) and retainer (3), first split tractor between engine and transmission as outlined in paragraph 93. Remove clutch release bearing (16), retainer (15), fork (13) and release shaft (12). Remove brake cross shaft, if so equipped. Remove screws and pull pto input shaft (8) and retainer (3) as a unit from transmission housing.

To disassemble, remove snap ring (6) from rear of retainer, then push input shaft and bearing (5) out of housing toward rear. Bearing can be removed from input shaft after removing snap ring (4). Remove oil seals (1 and 9) and needle bearing (2) only if new parts are to be installed.

Lubricate all seals and bearings with petroleum jelly before assembling. Use special tool (MF 315A or equivalent) to install new needle bearing (2). Install seal (9) with lip toward rear, using special tool (MF 421 or equivalent). Install ball bearing (5) on shaft with shield toward rear, then install snap ring (4). Use special tool (MF255B or equivalent) to install seal (1) with lip toward rear. Use special sleeve (KMF 1004 or equivalent) to protect seal (1), slide input shaft into the retainer housing, then install snap ring (6). Install "O" ring (7) and install housing and shaft assembly over transmission input shaft (11) into housing bore. Coat threads of retaining screws with "Hylomar" or equivalent sealer and tighten to 54-61 N•m (40-45 ft.-lbs.) torque. Remainder of assembly is reverse of disassembly.

PTO LOWER SHAFT FRONT BEARING AND RETAINER

Models With Eight-Speed Synchromesh Transmission

118. The pto lower shaft front bearing (41—Fig. 138) can be removed and installed after separating tractor between the rear of the engine and front of transmission housing. Removal of the lower driven gear (45) and shaft (46) require the additional steps of separating at rear of transmission housing and removal of the transmission top cover and shift rails.

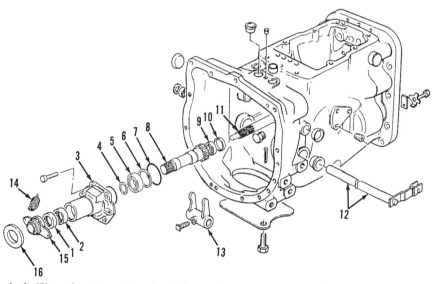

Fig. 137—Pto input shaft (8) and retainer housing (3) can be removed as a unit from front of transmission housing. Transmission input shaft (11) is removed rearward from transmission housing.

1. Seal	5. Ball bearing	9. Oil seal	13. Clutch release lever
2. Needle bearing	6. Snap ring	10. Thrust washer	14. Return spring
3. Retainer housing	7. "O" ring	11. Transmission input shaft	15. Release housing
4. Snap ring	8. Pto input shaft	12. Clutch lever and shaft	16. Release bearing

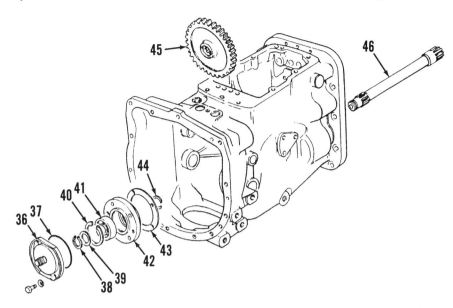

Fig. 138—Exploded view of the pto lower shaft and bearings.

36. Cover
37. "O" ring
38. Snap ring
39. Washer
40. Snap ring
41. Ball bearing
42. Retainer
43. Gasket
44. Snap ring
45. Pto lower gear
46. Pto lower shaft

The following procedure describes only service to the front bearing and bearing retainer (42) after splitting as outlined in paragraph 93.

Remove clutch release bearing (16—Fig. 137), retainer (15), fork (13) and release shaft (12). Unbolt and remove cover (36—Fig. 138), then remove snap ring (38) and washer (39). Thread a ⅜ inch UNC cap screw, 75 mm (3 inches) long into each of the two threaded holes of retainer (42). Tighten the two jackscrews to push retainer (42) and bearing (41) forward off shaft (46). Bearing can be pressed from retainer after removing snap ring (40). Remove "O" ring (37) and gasket (43), then clean all parts, thoroughly.

Coat gasket (43) and threads of retaining screws lightly with "Hylomar" or equivalent sealer before installing. End of shaft (46) is threaded to accept a puller screw to pull bearing (41) onto shaft until washer (39) and snap ring (38) can be installed. Align holes in retainer (42) with holes in transmission housing, install "O" ring (37) and cover (36). Tighten retaining screws to 54-61 N·m (40-45 ft.-lbs.) torque. Remainder of assembly is reverse of disassembly procedure.

TRANSMISSION INPUT SHAFT

Models With Eight-Speed Synchromesh Transmission

119. To remove the transmission input shaft (11—Fig. 139), first remove the transmission as outlined

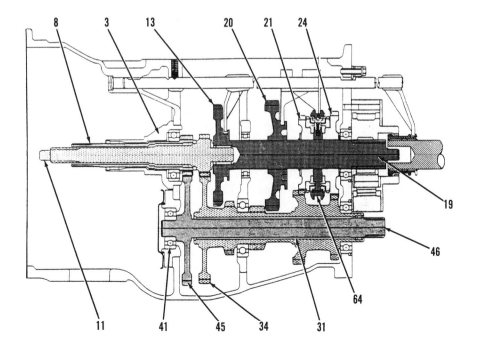

Fig. 139—Cross section of eight-speed transmission with third and fourth speeds of main gear box synchronized.

in paragraph 113. Remove the top cover and shift levers as outlined in paragraph 114, planetary unit as outlined in paragraph 116 and the shift rails and forks as outlined in paragraph 115. Refer to paragraph 118 and remove the pto lower shaft front bearing and retainer. Remove snap ring (44—Fig. 138), pull shaft (46) to rear and allow gear (45) to fall to bottom of housing. Move input shaft (11—Fig. 140) and gear (13) forward, then withdraw gear (13) out top. Transmission input shaft (11) can be removed from same top opening after gear (13) is out.

Reinstall input shaft by reversing removal procedures. Adjust position of shift fork in synchronizer collar as described in paragraph 115.

MAINSHAFT (OUTPUT SHAFT) AND GEARS

Models With Eight-Speed Synchromesh Transmission

120. To remove the mainshaft (19—Fig. 139), first remove the transmission as outlined in paragraph 113. Remove the top cover and shift levers as outlined in paragraph 114, planetary unit as outlined in paragraph 116 and the shift rails and forks as outlined in paragraph 115. Refer to paragraph 118 and remove the pto input shaft and retainer, remove the pto lower shaft front bearing and retainer as outlined in paragraph 118, then remove the transmission input shaft as outlined in paragraph 119.

Remove spacer (12—Fig. 140), lift snap ring (59) from groove and move snap ring forward on shaft. Push mainshaft (19) to rear and remove gear (13). Remove snap ring (14), then remove bearing (15) and snap ring (16). Continue to move mainshaft toward rear while working snap ring (59) forward on shaft. Withdraw mainshaft toward rear while lifting gears (20, 21 and 24) and synchronizer assembly out top opening. Rear bearing (22) can be removed if necessary after removing snap ring (23).

Synchronizer parts (64 through 67—Fig. 141) should be removed as an assembly and separated only if necessary. Wrap synchronizer assembly with cloth to catch poppet parts (65 and 66) when coupling (64) is removed from hub (67). Clean, inspect and renew any components that are damaged or worn excessively. Place synchronizer cones (72 and 73) and synchronizer rings (63 and 68) on respective gears (21 and 24). Make sure that each ring is seated squarely on taper of cone. Measure clearance between ring and gear at several places around gear, using a feeler gauge. Install new synchronizer ring (63 or 68) if clearance is less than 0.5 mm (0.020 inch). This clearance can be checked with gearbox assembled, but minimum clearance is increased to 0.8 mm (0.030 inch).

When reassembling the synchronizer shift components, align centralizing holes in shift collar (64) and hub (67). Special tools (MF 415 and MS 550) are available to assist holding the three poppets (65 and 66) into hub (67) while installing coupling (64).

Reassemble by reversing removal procedure. Be sure that sleeves (61 and 70) match respective gears (21 and 24) in length. Refer to Fig. 139 for cross section of assembled transmission. Coat gasket (43—Fig. 138) lightly with "Hylomar" or equivalent sealer before installing. Coat threads of screws attaching cover (36) and screws attaching retainer housing (3—Fig. 137) lightly with "Hylomar" or equivalent sealer before tightening to 54-61 N•m (40-45 ft.-lbs.) torque.

COUNTERSHAFT (LAYSHAFT) AND GEARS

Models With Eight-Speed Synchromesh Transmission

121. The mainshaft must be removed as outlined in paragraph 120 before countershaft. Remove snap ring (25—Fig. 140) from rear, bump shaft forward enough to remove snap ring (29) from groove and move snap ring forward to unsplined part of countershaft. Remove snap ring (35), then remove gears (33 and 34) while driving countershaft toward rear. After gears (33 and 34) are lifted from housing, drive countershaft toward front while removing gears (27 and 28).

Reassemble by reversing removal procedure. Snap ring (29) should be around shaft in the unsplined area and front bearing (32) should be in place on shaft before installing shaft in housing. Refer to Fig. 139 for cross section of assembled transmission.

REVERSE IDLER GEAR AND SHAFT

Models With Eight-Speed Synchromesh Transmission

122. A dummy shaft should be fabricated to hold parts in place while both removing and installing the reverse idler shaft. Dummy shaft should be 25 mm (1 inch) diameter and 55 mm (2 3/16 inches) long.

To remove the reverse idler, first remove the mainshaft and countershaft assemblies as outlined in paragraphs 120 and 121. Release locking washer, then remove locking screw and clip (58—Fig. 140). Slide dummy shaft in from front, pushing reverse idler shaft (57) out toward rear. Dummy shaft should be used to hold needle rollers and washers in place while both removing and installing the reverse idler. Be sure to remove all 56 of the needle rollers.

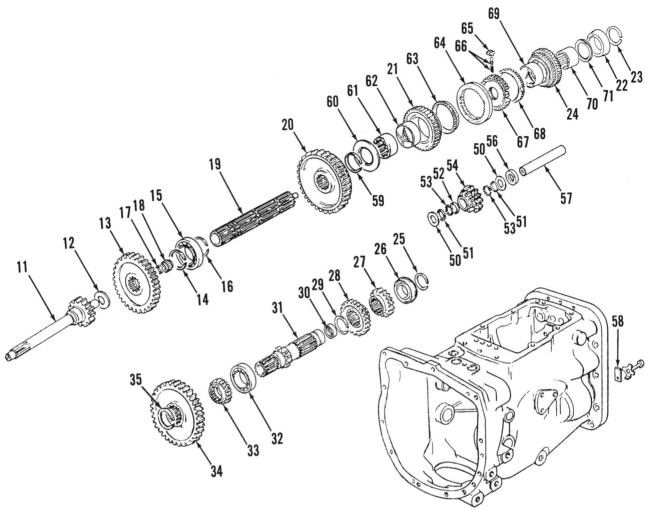

Fig. 140—Exploded view of eight-speed synchronized transmission main shaft and countershaft.

11. Transmission input shaft
12. Spacer
13. Gear (2nd)
14. Snap ring
15. Ball bearing
16. Snap ring
17. Spacer
18. Roller bearing
19. Mainshaft
20. Gear (1st & Reverse)
21. Gear (4th) and
    synchronizer cone
22. Ball bearing

23. Snap ring
24. Gear (3rd) and
    synchronizer cone
25. Snap ring
26. Ball bearing
27. Gear (3rd)
28. Gear (4th)
29. Snap ring
30. Roller bearing
31. Countershaft & 1st gear
32. Ball bearing
33. Gear (2nd)

34. Constant mesh gear
35. Snap ring
50. Thrust washers
51. Washers
52. Center spacer
53. Needle rollers
54. Gear (Reverse idler)
56. Spacer
57. Idler shaft
58. Retaining clip
59. Snap ring
60. Washer

61. Sleeve
62. Bushing
63. Synchronizer ring
    (Same as 68)
64. Shifting coupler
65. Pressure blocks (3 used)
66. Ball pins & springs (3 used)
67. Synchronizer hub
68. Synchronizer ring
    (Same as 63)
69. Bushing
70. Sleeve
71. Washer

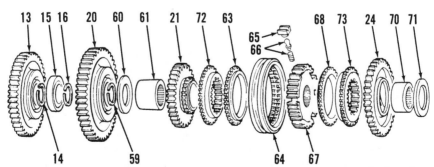

Fig. 141—Exploded view of synchronizer parts for third and fourth main gearbox speeds. Synchronizer cones (72 and 73) are not separated from gears in Fig. 140. Refer to Fig. 140 for legend.

Assemble needle rollers, washers and spacer in reverse idler using petroleum jelly and the dummy shaft. Assemble center spacer (52), one path of 28 needle rollers (53), one end washer (51) and one thrust washer (50), around dummy shaft, in one end of reverse idler gear. Assemble the second path of 28 needle rollers, the other end washer (51) and the remaining thrust washer (50). Center the dummy shaft so that it holds both thrust washers in place, then position assembly in housing and install idler shaft (57). Install clip (58), lock plate and retaining screw. Tighten retaining screw and lock by bending tab of lock plate against flat of screw. Complete reassembly by reversing disassembly procedure.

TRANSMISSION (TWELVE-SPEED SYNCHROMESH)

362, 365, 375, 390, 390T and 398 Models

Models 362, 365, 375, 390, 390T and 398 are available with a transmission providing twelve forward speeds and four reverse speeds. The twelve forward speeds and four reverse speeds are possible by a manually shifted, high/low range, located ahead of the main gearbox, a main gearbox with three forward speeds and one reverse speed, and finally by a two speed planetary unit which compounds the number of speeds already provided. The high/low range and the top two forward gears of the main gearbox are synchronized. Refer to Fig. 142 for shift pattern of twelve-speed synchronized transmission models.

LUBRICATION

Models With Twelve-Speed Synchromesh Transmission

123. The transmission, rear differential and hydraulic system share common fluid which is contained in the transmission and center housing. Recommended lubricant is Massey-Ferguson Super 500 multi-use 10W-30 oil, Massey-Ferguson Permatran oil or equivalent transmission/hydraulic oil. Capacity is 47.4 L (12.5 gal.) if equipped with spacer or four-wheel-drive transfer box between transmission and rear axle center housing; 43.4 L (11.5 gal.) if not equipped with transfer box or spacer. Tractors are equipped with dipstick (Fig. 143 or Fig. 144) with marks indicating maximum and minimum oil levels. Level should be maintained at maximum level if

Fig. 142—Shift pattern for twelve-speed synchronized transmission. Only third and fourth speeds of the main transmission and high/low (indicated by tortoise and hare/rabbit) are synchronized.

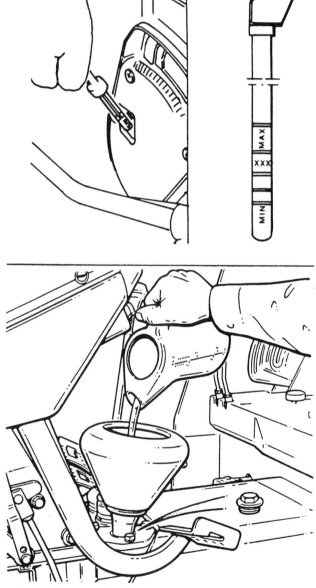

Fig. 143—Transmission oil level of tractors without cab is measured by dipstick located in the side cover as shown. Refer to Fig. 144 for models with cab.

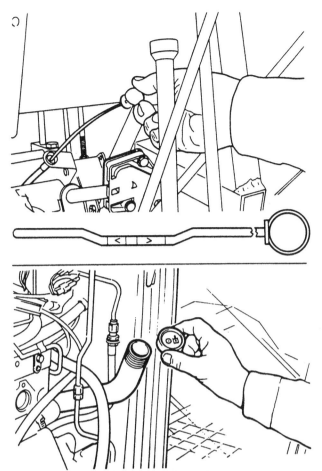

Fig. 144—Transmission oil level of tractors with cab is measured by dipstick that points toward rear as shown.

tractor is operated on hilly terrain or if using implements which require large quantities of oil.

TRACTOR REAR SPLIT

Models With Twelve-Speed Synchromesh Transmission

124. To separate the rear of the transmission from the rear axle center housing (Fig. 145), refer to paragraph 99 if tractor does not have cab or to paragraph 100 if equipped with cab.

TRANSMISSION REMOVAL

Models With Twelve-Speed Synchromesh Transmission

125. To remove the transmission, first separate the rear of the transmission from rear axle center housing as outlined in paragraph 99 if tractor does not have cab or to paragraph 100 if equipped with cab. If not equipped with cab, unbolt and support or remove the steering and instrument console. On all models, support the engine and transmission housing separately. Remove screws and stud nuts, then separate transmission from engine.

Reassemble transmission to engine by reversing the splitting procedure and observe the following. Install a M12 alignment stud approximately 100 mm (4 inches) long in each side of the transmission hous-

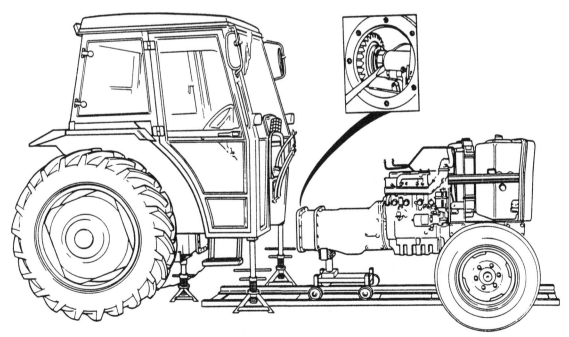

Fig. 145—View of tractor with cab separated between the four-wheel-drive transfer case and the rear axle center housing. Special splitting stand and track is shown. Support and splitting procedures are similar for models without cab and without four-wheel drive.

ing to assist in aligning engine to transmission. Turn flywheel to align clutch plate splines with transmission and pto input shaft splines. Install retaining screws and nuts after engine and transmission housing flanges are completely together. Tighten retaining screws and nuts to 115 N·m (85 ft.-lbs.) torque.

Refer to paragraph 99 or 100 and reconnect transmission to rear axle center housing. Tighten all screws retaining the transmission housing to spacer housing, transfer case or rear axle center housing to 112 N·m (83 ft.-lbs.) torque, beginning at top center and progressing in a clockwise direction (as viewed from tractor rear) two times around the mating flange. On four-wheel-drive models, coat threads of drive shaft retaining screws with "Loctite 270," then attach drive shaft flanges and tighten screws to 55-75 N·m (40-55 ft.-lbs.) torque. On all models, check and adjust clutch linkage as outlined in paragraph 92.

TOP COVER

Models With Twelve-Speed Synchromesh Transmission

126. The shift levers and the small shift tower can be removed after first removing floor mat and access panels. Installation of the shift levers and tower is easier if transmission is shifted into neutral before removing. Shift tower (11—Fig. 146) is attached to the top cover (18) by five screws.

The main shift lever (6) can be removed after removing the tower assembly, cover (3), snap ring (9), spring (8) and pin (10). The high/low shift lever (13) and over/under shift lever (19) can be removed after removing cover (14) and pin (16). Refer to Fig. 147 for cross section of shift levers.

The transmission top cover (18—Fig. 146) can be unbolted and lifted from transmission after first removing the shift levers and tower assembly. On models with cab, it is necessary to split rear of transmission from rear axle center housing as outlined in paragraph 100 or remove the cab assembly. Models without cab must have instrument console and other interfering parts removed before transmission top cover. The reason for removing top cover will determine what additional procedures are necessary.

"Loctite 515 Instant Gasket" or equivalent should be used to seal tower to top cover. Tighten retaining screws to 50-70 N·m (37-52 ft.-lbs.) torque.

SHIFTER RAILS AND FORKS

Models With Twelve-Speed Synchromesh Transmission

127. To remove the shifter rails and forks, first refer to paragraph 99 or paragraph 100 and separate trac-

tor between rear of transmission and front of axle center housing. Refer to paragraph 126 and remove the shift lever and tower assembly and transmission top cover. Models without cab must have instrument and steering console removed before top cover can be unbolted and removed.

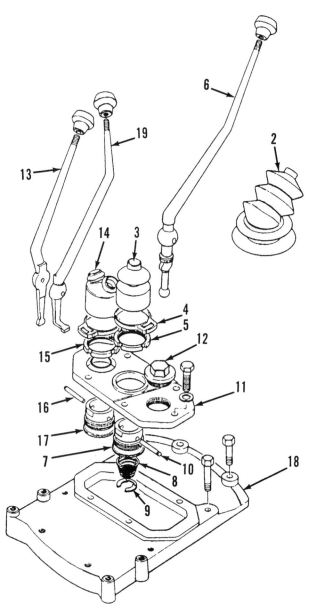

Fig. 146—Exploded view of transmission top cover, tower cover and shift levers typical of models with twelve-speed transmission.

2. Boot	
3. Cover	12. Fill plug
4. Retainer clip	13. High/low shift lever
5. Retaining nut	14. Cover
6. Main shift lever	15. Retaining nut
7. Sleeve	16. Pin
8. Spring	17. Sleeve
9. Snap ring	18. Top cover
10. Pin	19. Over/Underdrive
11. Tower cover	shift lever

On all models, remove the neutral safety start switch (N—Fig. 149). Remove safety wire and set screws (5—Fig. 148) and lift detent springs and plungers (4) from bores in housing. Unbolt and remove interlock parts (12, 13, 14 and 15—Fig. 149). Slide rail out of housing toward rear and remove gate and shift fork through top. Twist rails if necessary while withdrawing. Fork (8—Fig. 148) can not be removed until transmission input shaft and gears are removed as outlined in paragraph 131.

When assembling, lubricate rails and make sure that rails do not have burrs, especially around the set screw locations, that would prevent easy and smooth installation. Each of the four shift rails is different from the other rails. Tighten screws retaining the interlock retainer (15—Fig. 148) to 40-47 N·m (30-35 ft.-lbs.) torque and the seven set screws (5) to 34-52 N·m (25-38 ft.-lbs.) torque. Install safety wire through set screws (5) after tightening to prevent loosening. Adjust position of shift fork (18) as follows.

Use special tools (MF 414/2 and MF 414/3 or equivalent) to hold shift rails (6 and 16—Fig. 148 and Fig. 149) in neutral detent position. Insert centralizing tool (MF 414/1 or equivalent) into holes in shift forks (8 and 18—Fig. 148) and into matching hole of sliding synchronizing coupling. Tighten each set screws (5S), while turning centralizing tool, to hold the fork in this center position. Screws (5 and 5S) should be tightened to 34-52 N·m (25-38 ft.-lbs.) torque and locked in position with safety wire. Shift all gears, rails and gates to neutral after checking to make sure that all shift gates align. Shift into each gear to make sure that each can be correctly engaged. Install detent plungers and springs (4—Fig. 149). Remainder of assembly is the reverse of disassembly procedure.

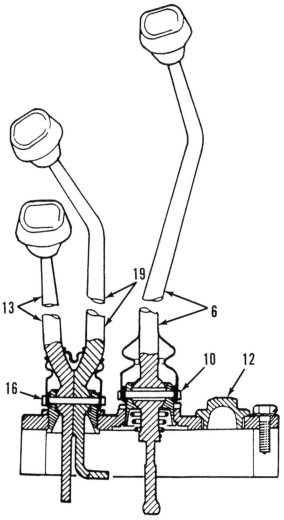

Fig. 147—Cross section of shift tower used on tractors with twelve-speed transmission. Refer to Fig. 146 for legend.

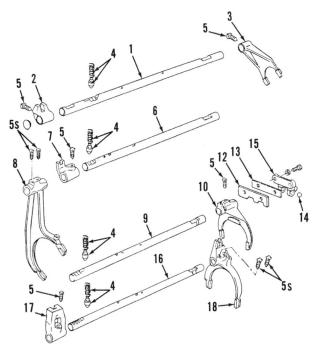

Fig. 148—Exploded view of shift rails and forks for twelve-speed transmission with synchronized second and third speeds in the main transmission and synchronized over/under drive.

1. Shift rail (High/Low)	9. Shift rail (1st/Rev.)
2. Shift gate	10. Shift fork
3. Shift fork	12. Plate
4. Detent springs	13. Gasket
and plungers	14. Interlock plunger
5. Retaining screws	& balls
6. Shift rail (Over/	15. Retainer
Underdrive)	16. Shift rail (2nd/3rd)
7. Shift gate	17. Shift gate
8. Shift fork	18. Shift fork

109

PLANETARY UNIT

Models With Twelve-Speed Synchromesh Transmission

128. To remove the planetary unit, first refer to paragraph 99 or paragraph 100 and separate tractor between rear of transmission and front of axle center housing. Remove set screw (5—Fig. 148) from fork (3), then remove the fork and coupler from rear of planetary. Remove the four retaining cap screws and withdraw rear cover (13—Fig. 150 or Fig. 151), rear thrust ring (6) and planet carrier (7). Pry planetary ring gear (3) with dowels from rear of housing. Remove planetary front cover (2) and shim (1).

To disassemble planet carrier (7), first remove snap ring (5—Fig. 151), if used. Press planet pins (8—Fig. 150 or Fig. 151) forward out of carrier. Remove planet gears (10) with needle rollers (12) and thrust washers (9). Splined sleeve (15) is retained in hub of carrier (7) by retaining ring (16).

When reassembling, be sure to account for all of the needle rollers (12). Each pinion contains two paths of rollers separated by a washer (11). Each roller path contains 27 rollers (12—Fig. 150) on 362 and 365 models. Each roller path contains 16 rollers (12—Fig. 151) on 375, 390, 390T and 398 models. Use petroleum jelly (not heavy grease) to hold rollers in place

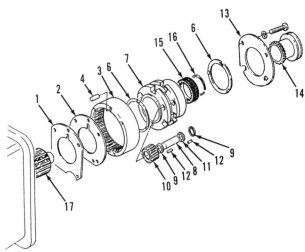

Fig. 150—Exploded view of high/low range planetary unit used on 362 and 365 models. Refer to Fig. 151 for other models with twelve-speed synchronized transmission.

1. Front shim	10. Pinion
2. Front plate	11. Spacer washer
3. Ring gear	12. Needle rollers (27/path)
4. Dowel	13. Rear plate
6. Thrust washer	14. Shift coupler
7. Planet carrier	15. Splined sleeve
8. Pinion shaft	16. Retaining ring
9. Side washer	17. Transmission mainshaft

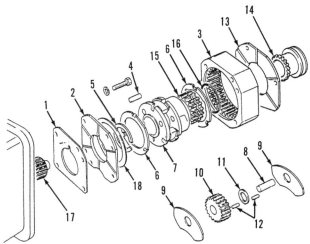

Fig. 151—Exploded view of high/low range planetary unit used on 375, 390, 390T and 398 models. Refer to Fig. 150 for planetary unit used on other models with twelve-speed synchronized transmission.

1. Front shim	10. Pinion
2. Front plate	11. Spacer washer
3. Ring gear	12. Needle rollers (16/path)
4. Dowel	13. Rear plate
5. Snap ring	14. Shift coupler
6. Thrust washer	15. Splined sleeve
7. Planet carrier	16. Retaining ring
8. Pinion shaft	17. Transmission mainshaft
9. Side washer	18. Belleville washer

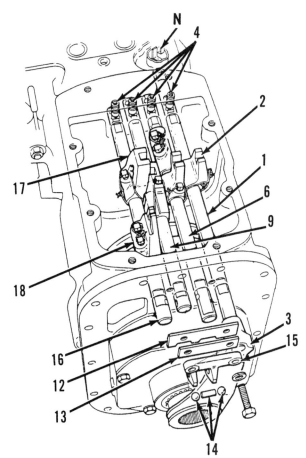

Fig. 149—Drawing of shift rails, installed. Notice position of notches in plate (12) and location of neutral start switch (N). Plunger of interlock (14) is located in rail (9) and balls are located in retainer (15). Refer to Fig. 148 for legend.

while pressing planet pins into carrier. Gap in snap ring (5—Fig. 151) should be located between planet pins.

Concave side of Belleville washer (18—Fig. 151) should be toward rear. Be sure that front plate (2—Fig. 150 or Fig. 151) and rear plate (13) are positioned with grooved side toward planet carrier. Slots in front plate and shim (1) must be aligned and toward top when installed on transmission. Use petroleum jelly to hold thrust rings (6) in place. Make certain that tangs of thrust rings engage notches in pinion carrier and that brass side faces away from carrier. On 362 and 365 models, cutaway section of rear plate (13—Fig. 150) should be positioned at the lower, **right** corner of planetary as shown. A lockwasher is used only on mounting screw at corner of cutaway. On 375, 390, 390T and 398 models, the slanted section of rear plate (13—Fig. 151) should be positioned at lower, **left** corner of planetary. On all models, tighten mounting screws to 40-47 N·m (30-35 ft.-lbs.) torque. Complete assembly procedure by reversing disassembly.

PTO INPUT SHAFT AND RETAINER HOUSING

Models With Twelve-Speed Synchromesh Transmission

129. To remove the pto input shaft (8—Fig. 152) and retainer (3), first split tractor between engine and transmission as outlined in paragraph 93. Remove clutch release bearing (16), retainer (15), fork (13) and release shaft (12). Remove brake cross shaft, if so equipped. Remove screws and pull pto input shaft (8) and retainer (3) as a unit from transmission housing.

To disassemble, remove snap ring (6) from rear of retainer, then push input shaft and bearing (5) out of housing toward rear. Bearing can be removed from input shaft after removing snap ring (4). Remove oil seals (1 and 9) and needle bearing (2) only if new parts are to be installed.

Lubricate all seals and bearings with petroleum jelly before assembling. Use special tool (MF 315A or equivalent) to install new needle bearing (2). Install seal (9) with lip toward rear, using special tool (MF 421 or equivalent). Install ball bearing (5) on shaft with shield toward rear, then install snap ring (4). Use special tool (MF255B or equivalent) to install seal (1) with lip toward rear. Use special sleeve (KMF 1004 or equivalent) to protect seal (1), slide input shaft into the retainer housing, then install snap ring (6). Install "O" ring (7) and install housing and shaft assembly over transmission input shaft (11) into housing bore. Coat threads of retaining screws with "Hylomar" or equivalent sealer and tighten to 54-61 N·m (40-45 ft.-lbs.) torque. Remainder of assembly is reverse of disassembly.

PTO LOWER SHAFT FRONT BEARING AND RETAINER

Models With Twelve-Speed Synchromesh Transmission

130. The pto lower shaft front bearing (41—Fig. 153) can be removed and installed after separating tractor between the rear of the engine and front of transmission housing. Removal of the lower driven gear (45) and shaft (46) require the additional steps of separating at rear of transmission housing and removal of the transmission top cover and shift rails. The following procedure describes only service to the

Fig. 152—Pto input shaft (8) and retainer housing (3) can be removed as a unit from front of transmission housing. Transmission input shaft (11) is removed rearward from transmission housing.

1. Seal
2. Needle bearing
3. Retainer housing
4. Snap ring
5. Ball bearing
6. Snap ring
7. "O" ring
8. Pto input shaft
9. Oil seal
10. Thrust washer
11. Transmission input shaft
12. Clutch lever and shaft
13. Clutch release lever
14. Return spring
15. Release housing
16. Release bearing

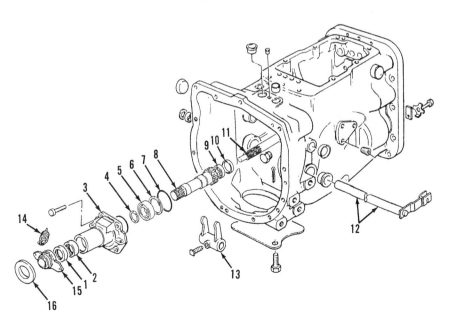

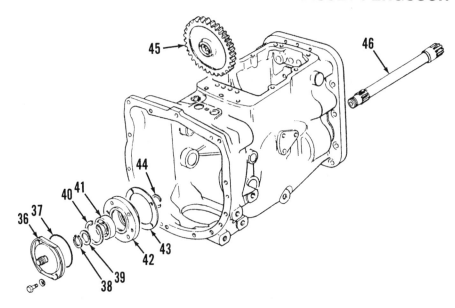

Fig. 153—Exploded view of the pto lower shaft and bearings.

36. Cover
37. "O" ring
38. Snap ring
39. Washer
40. Snap ring
41. Ball bearing
42. Retainer
43. Gasket
44. Snap ring
45. Pto lower gear
46. Pto lower shaft

front bearing and bearing retainer (42) after splitting as outlined in paragraph 93.

Remove clutch release bearing (16—Fig. 152), retainer (15), fork (13) and release shaft (12). Unbolt and remove cover (36—Fig. 153), then remove snap ring (38) and washer (39). Thread a ⅜ inch UNC cap screw, 75 mm (3 inches) long into each of the two threaded holes of retainer (42). Tighten the two jackscrews to push retainer (42) and bearing (41) forward off shaft (46). Bearing can be pressed from retainer after removing snap ring (40). Remove "O" ring (37) and gasket (43), then clean all parts, thoroughly.

Coat gasket (43) and threads of retaining screws lightly with "Hylomar" or equivalent sealer before installing. End of shaft (46) is threaded to accept a puller screw to pull bearing (41) onto shaft until washer (39) and snap ring (38) can be installed. Align holes in retainer (42) with holes in transmission housing, install "O" ring (37) and cover (36). Tighten retaining screws to 54-61 N•m (40-45 ft.-lbs.) torque. Remainder of assembly is reverse of disassembly procedure.

TRANSMISSION INPUT SHAFT

Models With Twelve-Speed Synchromesh Transmission

131. To remove the transmission input shaft (11—Fig. 154), first remove the transmission as outlined in paragraph 125. Remove the top cover and shift levers as outlined in paragraph 126, planetary unit as outlined in paragraph 128 and the shift rails and forks as outlined in paragraph 127. Refer to paragraph 129 and remove the pto input shaft and retainer. Slide input shaft (11—Fig. 155) forward, then withdraw gears (81 and 82) out top. Over/Under shift

fork (8—Fig. 148) can be removed after removing input gears.

Reinstall input shaft by reversing removal procedures. Refer to paragraph 127 for adjusting neutral position of forks on shift shafts.

MAINSHAFT (OUTPUT SHAFT) AND GEARS

Models With Twelve-Speed Synchromesh Transmission

132. To remove the mainshaft (19—Fig. 154), first remove the transmission as outlined in paragraph 125. Remove the top cover and shift levers as outlined in paragraph 126, planetary unit as outlined in paragraph 128 and the shift rails and forks as outlined in paragraph 127. Refer to paragraph 128 and remove the pto input shaft and retainer, then remove the transmission input shaft as outlined in paragraph 131.

Remove spacer (83—Fig. 155), lift snap ring (14) from groove, push mainshaft (19) to rear and remove gear (13). Continue to move mainshaft toward rear while lifting synchronizer assembly (91 and 92), bearing (93), gears (20, 21 and 24) and synchronizer assembly (63 through 68) out top opening. Rear bearing (22) can be removed if necessary after removing snap ring (23).

Synchronizer parts (64 through 67—Fig. 156) should be removed as an assembly and separated only if necessary. Wrap synchronizer assembly with a cloth to catch poppet parts (65 and 66) when coupling (64) is removed from hub (67). Clean, inspect and renew any components that are damaged or worn excessively. Place synchronizer cones (72 and 73—Fig. 155) and synchronizer rings (63 and 68) on re-

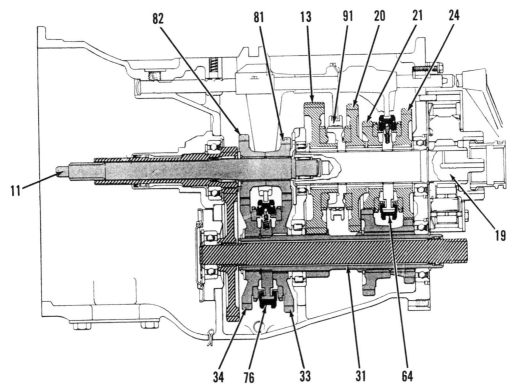

Fig. 154—Cross section of twelve-speed transmission.

spective gears (21 and 24). Make sure that each ring is seated squarely on taper of cone. Measure clearance between ring and gear at several places around gear, using a feeler gauge. Install new synchronizer ring (63 or 68) if clearance is less than 0.5 mm (0.020 inch). This clearance can be checked with gearbox assembled, but minimum clearance is increased to 0.8 mm (0.030 inch).

When reassembling the synchronizer shift components, align centralizing holes in shift collar (64) and hub (67). Special tools (MF 415 and MS 550) are available to assist holding the three poppets (65 and 66) into hub (67) while installing coupling (64).

Make sure that Low (Tortoise) and High (Hare) shift fork is in place in shift collar before installing the mainshaft.

Reassemble mainshaft (19—Fig. 155), snap rings (14 and 23), bearings (15 and 22) and all parts between snap rings (14 and 23), outside of gearbox housing. Be sure that sleeves (61 and 70) match respective gears (21 and 24) in length. Refer to Fig. 157 for cross section of assembled mainshaft. Measure clearance (G) between front snap ring (14) and inner race of bearing (15). Select spacer (17—Fig. 155 and Fig. 157) of proper thickness to limit gap (G) to 0.080-0.300 mm (0.003-0.012 inch). Spacer (17) is available in four thicknesses of 4.14-4.19, 4.39-4.44, 4.62-4.67 and 4.85-4.90 mm (0.163-0.165, 0.173-0.175, 0.182-0.184 and 0.191-0.193 inch). Reassemble mainshaft and components inside housing using the preselected spacer (17). Shorter sleeve (70—Fig. 155)

is in 2nd gear (24) and longer sleeve (61) is in 3rd and reverse gears (20 and 72). Flat side of spacer (17) should be toward gear (13).

Remainder of assembly is reverse of disassembly procedure. Coat threads of screws attaching retainer housing (3—Fig. 152) lightly with "Hylomar" or equivalent sealer before tightening to 54-61 N·m (40-45 ft.-lbs.) torque.

COUNTERSHAFT (LAYSHAFT) AND GEARS

Models With Twelve-Speed Synchromesh Transmission

133. The mainshaft must be removed as outlined in paragraph 132 and pto lower shaft should be removed as outlined in paragraph 130, before removing countershaft. Remove snap ring (25—Fig. 155) from rear, bump shaft forward enough to remove snap ring (29) from groove and move snap ring forward to unsplined part of countershaft. Remove snap ring (35) and thrust washer (84). Slide gear (34) forward, remove snap ring (86) from groove, then slide snap ring forward as far as possible. Slide shaft (31) toward rear as far as possible and remove gear (34). Remove snap ring (86) and synchronizer ring (75), then remove synchronizer assembly (76 through 79) as a unit. Continue sliding shaft (31) toward rear and remove synchronizer ring (80), gear (33) and washer

(88). Slide shaft (31) toward front and remove gears (27 and 28).

Synchronizer parts (76 through 79—Fig. 156) should be removed as an assembly and separated only if necessary. Wrap synchronizer assembly with a cloth to catch poppet parts (77 and 78) when coupling (76) is removed from hub (79). Clean, inspect and renew any components that are damaged or worn excessively. Place synchronizer rings (75 and 80—Fig. 155) on respective gears (33 and 34). Make sure that each ring is seated squarely on taper of cone. Measure clearance between ring and gear at several places around gear, using a feeler gauge. Install new synchronizer ring if clearance is less than 0.5 mm

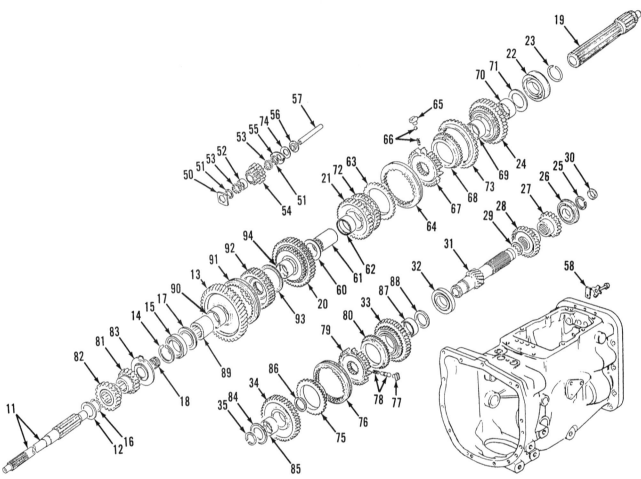

Fig. 155—Exploded view of twelve-speed transmission. Synchronizer (63 through 68) is made up of same components as synchronizer (75 through 80).

11. Transmission input shaft	28. Gear (3rd)	62. Bushing
12. Spacer	29. Snap ring	63. Synchronizer ring (Same as 68)
13. Gear (1st)	30. Roller bearing	64. Shifting coupler
14. Snap ring	31. Countershaft & 1st gear	65. Pressure blocks (3 used)
15. Bearing	32. Ball bearing	66. Ball pins & springs (3 used)
16. Snap ring	33. Gear (Low/Tortoise)	67. Synchronizer hub
17. Spacer (variable thicknesses)	34. Gear (High/Hare)	68. Synchronizer ring (Same as 63)
18. Roller bearing	35. Snap ring	69. Bushing
19. Mainshaft	50. Thrust washer	70. Sleeve (short)
20. Gear (Reverse)	51. Washers	71. Washer
21. Gear (3rd)	52. Center spacer	72. Synchronizer cone
22. Ball bearing	53. Needle rollers	73. Synchronizer cone
23. Snap ring	54. Gear (Reverse idler)	74. Washer
24. Gear (2nd)	55. Friction disc	75. Synchronizer ring (Same as 80)
25. Snap ring	56. Spacer	76. Shifting coupler
26. Ball bearing	57. Idler shaft	77. Pressure blocks (3 used)
27. Gear (2nd)	58. Retaining clip	78. Ball pins & springs (3 used)
	60. Washer	79. Synchronizer hub
	61. Sleeve (longer)	80. Synchronizer ring (Same as 75)
		81. Gear (Low/Tortoise)
		82. Gear (High/Hare)
		83. Spacer
		84. Thrust washer
		85. Bushing
		86. Snap ring
		87. Bushing
		88. Thrust washer
		89. Sleeve (medium)
		90. Bushing
		91. Shifting coupler
		92. Hub
		93. Thrust bearing
		94. Bushing

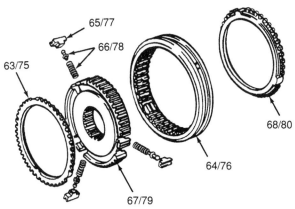

Fig. 156—Exploded view of synchronizer parts for second and third main gearbox speeds and Low/Tortoise and High/Hare speeds. Refer to Fig. 155 for legend.

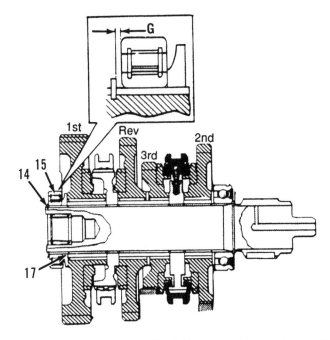

Fig. 157—The mainshaft should be assembled outside of transmission housing to determine correct thickness of thrust washer (17) as determined by gap (G).

(0.020 inch). This clearance can be checked with gearbox assembled, but minimum clearance is increased to 0.8 mm (0.030 inch).

When reassembling the synchronizer shift components, align centralizing holes in shift collar (76) and hub (79). Special tools (MF 415 and MS 550) are available to assist holding the three poppets (77 and 78) into hub (79) while installing coupling (76).

Reassemble by reversing removal procedure. Always install new snap rings when reassembling. Position snap ring (29) around shaft in the unsplined area and install front bearing (32) on shaft before installing shaft in housing. Refer to Fig. 154 for cross section of assembled transmission. Refer to paragraph 132 for assembly of mainshaft and selection of thrust washer (17—Fig. 155 and Fig. 157).

Coat gasket (43—Fig. 153) lightly with "Hylomar" or equivalent sealer before installing. Coat threads of screws attaching cover (36) and screws attaching retainer housing (3—Fig. 152) lightly with "Hylomar" or equivalent sealer before tightening to 54-61 N·m (40-45 ft.-lbs.) torque.

REVERSE IDLER GEAR AND SHAFT

Models With Twelve-Speed Synchromesh Transmission

134. A dummy shaft should be fabricated to hold parts in place while both removing and installing the reverse idler shaft. Dummy shaft should be 25 mm (1 inch) diameter and 55 mm (2³⁄₁₆ inches) long.

To remove the reverse idler, first remove the mainshaft and countershaft assemblies as outlined in paragraphs 132 and 133. Release locking washer, then remove locking screw and clip (58—Fig. 155). Slide dummy shaft in from front, pushing reverse idler shaft (57) out toward rear. Dummy shaft should be used to hold needle rollers and washers in place while both removing and installing the reverse idler. Be sure to remove all 56 of the needle rollers.

Assemble needle rollers, washers and spacer in reverse idler using petroleum jelly and the dummy shaft. Assemble center spacer (52), one path of 28 needle rollers (53), one end washer (51) and one thrust washer (50), around dummy shaft, in one end of reverse idler gear. Assemble the second path of 28 needle rollers, the other end washer (51) and the remaining thrust washer (50). Center the dummy shaft so that it holds both thrust washers in place, then position assembly in housing and install idler shaft (57). Install clip (58), lock plate and retaining screw. Tighten retaining screw and lock by bending tab of lock plate against flat of screw. Complete reassembly by reversing disassembly procedure.

TRANSMISSION
(TWELVE-SPEED MULTI-POWER)

365, 375, 390 and 398 Models

Models 365, 375, 390 and 398 are available with a multi-power transmission providing twelve forward speeds and four reverse speeds. The transmission is divided into three sections. The front section consists of a hydraulic clutch operated, constant mesh, two speed, high/low gear reduction, located just ahead of the main gearbox. Second is a main gearbox with three forward speeds and one reverse speed. The top two forward gears of the main gearbox are synchronized. Finally a two speed planetary unit compounds the number of speeds already provided. Refer to Fig. 158 for shift pattern of twelve-speed multi-power transmission.

LUBRICATION

Models With Twelve-Speed Multi-Power Transmission

135. The transmission, rear differential and hydraulic system share common fluid which is contained in the transmission and center housing. Recommended lubricant is Massey-Ferguson Super 500 multi-use 10W-30 oil, Massey-Ferguson Permatran oil or equivalent transmission/hydraulic oil. Capacity is 47.4 L (12.5 gal.) if equipped with spacer or four-wheel-drive transfer box between transmission and rear axle center housing; 43.4 L (11.5 gal.) if not equipped with transfer box or spacer. Tractors are equipped with dipstick (Fig. 159 or Fig. 160) with marks indicating maximum and minimum oil levels. Level should be maintained at maximum level if tractor is operated on hilly terrain or if using implements which require large quantities of oil.

TRACTOR REAR SPLIT

Models With Twelve-Speed Multi-Power Transmission

136. To separate the rear of the transmission from the rear axle center housing, refer to paragraph 99 if

tractor does not have cab or to paragraph 100 if equipped with cab.

TRANSMISSION REMOVAL

Models With Twelve-Speed Multi-Power Transmission

137. To remove the transmission, first separate the rear of the transmission from rear axle center hous-

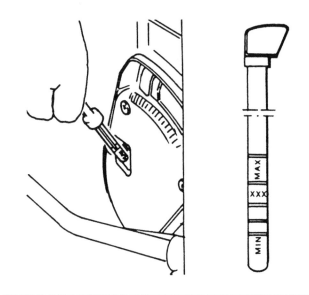

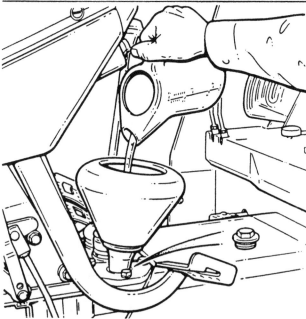

Fig. 159—Transmission oil level of tractors without cab is measured by dipstick located in the side cover as shown. Refer to Fig. 160 for models with cab.

Fig. 158—Shift pattern for twelve-speed multi-power transmission.

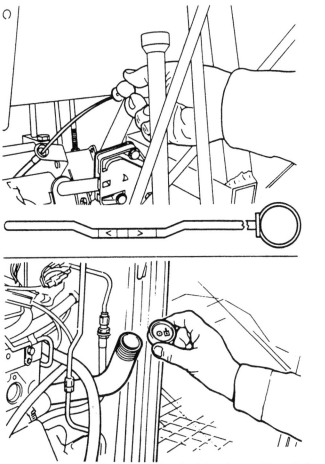

Fig. 160—Transmission oil level of tractors with cab is measured by dipstick that points toward rear as shown.

ing as outlined in paragraph 99 if tractor does not have cab or to paragraph 100 if equipped with cab. Refer to Fig. 161. If not equipped with cab, unbolt and support or remove the steering and instrument console. On all models, support the engine and transmission housing separately. Remove screws and stud nuts, then separate transmission from engine.

Reassemble transmission to engine by reversing the splitting procedure and observe the following. Install a M12 alignment stud approximately 100 mm (4 inches) long in each side of the transmission housing to assist in aligning engine to transmission. Turn flywheel to align clutch plate splines with transmission and pto input shaft splines. Install retaining screws and nuts after engine and transmission housing flanges are completely together. Tighten retaining screws and nuts to 115 N•m (85 ft.-lbs.) torque.

Refer to paragraph 99 or 100 and reconnect transmission to rear axle center housing. Tighten all screws retaining the transmission housing to spacer housing, transfer case or rear axle center housing to 112 N•m (83 ft.-lbs.) torque, beginning at top center and progressing in a clockwise direction (as viewed from tractor rear) two times around the mating flange. On four-wheel-drive models, coat threads of drive shaft retaining screws with "Loctite 270," then attach drive shaft flanges and tighten screws to 55-75 N•m (40-55 ft.-lbs.) torque. On all models, check and adjust clutch linkage as outlined in paragraph 92.

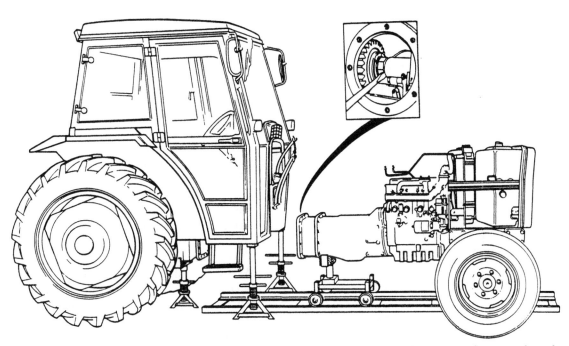

Fig. 161—View of tractor with cab separated between the transmission housing and the rear axle center housing. Special splitting stand and track is shown. Support and splitting procedures are similar for models without cab.

TOP COVER

Models With Twelve-Speed Multi-Power Transmission

138. The shift levers and the small shift tower can be removed after first removing floor mat and access panels. Installation of the shift levers and tower is easier if transmission is shifted into neutral before removing. Shift tower (11—Fig. 162) is attached to the top cover (18) by five screws.

The main shift lever (6) can be removed after removing the tower assembly, cover (3), snap ring (9), spring (8) and pin (10). The high/low shift lever can be removed after removing cover (14) and pin (16).

The transmission top cover (18) can be unbolted and lifted from transmission after first removing the shift levers and tower assembly. On models with cab, it is necessary to split rear of transmission from rear axle center housing as outlined in paragraph 100 or remove the cab assembly. Models without cab must have instrument console and other interfering parts removed before transmission top cover. The reason for removing top cover will determine what additional procedures are necessary.

"Loctite 515 Instant Gasket" or equivalent should be used to seal tower to top cover. Tighten retaining screws to 50-70 N·m (37-52 ft.-lbs.) torque.

SHIFTER RAILS AND FORKS

Models With Twelve-Speed Multi-Power Transmission

139. To remove the shifter rails and forks, first refer to paragraph 99 or paragraph 100 and separate tractor between rear of transmission and front of axle center housing. Refer to paragraph 138 and remove the shift lever and tower assembly and transmission top cover. Models without cab must have instrument and steering console removed before top cover can be unbolted and removed.

On all models, remove the neutral safety start switch (N—Fig. 163). Remove safety wire and set screws (5—Fig. 164) and lift detent springs and plungers (4) from bores in housing. Unbolt and remove interlock retainer (15) and plunger (14). Slide rails out of housing toward rear and remove gates and shift forks through top. Twist rails if necessary while withdrawing.

When assembling, lubricate rails and make sure that rails do not have burrs, especially around the set screw locations, that would prevent easy and smooth installation. Each of the three shift rails is different from the other rails. Tighten screws retaining the interlock retainer (15—Fig. 163) to 40-47 N·m (30-35 ft.-lbs.) torque and the five set screws (5) to 34-52 N·m

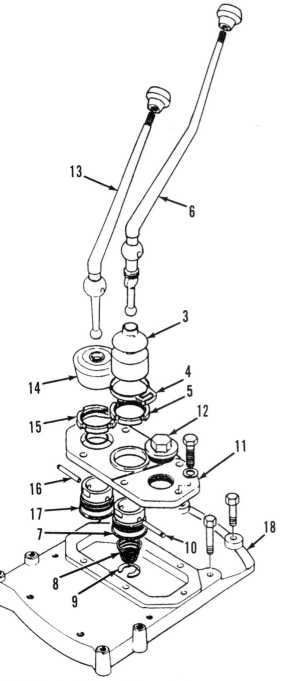

Fig. 162—Exploded view of transmission top cover, tower cover and shift levers typical of models with twelve-speed multi-power transmission.

3. Cover	11. Tower cover
4. Retainer clip	12. Fill plug
5. Retaining nut	13. High/low shift lever
6. Main shift lever	14. Cover
7. Sleeve	15. Retaining nut
8. Spring	16. Pin
9. Snap ring	17. Sleeve
10. Pin	18. Top cover

(25-38 ft.-lbs.) torque. Install safety wire through set screws (5) after tightening to prevent loosening. Adjust position of shift fork (18) as follows.

Use special tools (MF 414/2 and MF 414/3 or equivalent) to hold shift rail (16—Fig. 163 and Fig. 164) in neutral detent position. Insert centralizing tool (MF 414/1 or equivalent) into hole (H—Fig. 164) and into hole of sliding synchronizing coupling. Tighten each of the two set screws (5S), while turning centralizing tool, to hold the fork in this center position. Screws (5S) should also be tightened to 34-52 N.m (25-38 ft.-lbs.) torque and locked in position with safety wire. Shift all gears, rails and gates to neutral after checking to make sure that shift gates all align. Shift into each gear to make sure that each can be correctly engaged. Install detent plungers and springs (4—Fig. 163). Remainder of assembly is the reverse of disassembly procedure.

PLANETARY UNIT

Models With Twelve-Speed Multi-Power Transmission

140. To remove the planetary unit, first refer to paragraph 99 or paragraph 100 and separate tractor between rear of transmission and front of axle center housing. Remove set screw (5—Fig. 164) from fork (3), then remove the fork and coupler from rear of planetary. Remove the four retaining cap screws and withdraw rear cover (13—Fig. 165 or Fig. 166), rear thrust ring (6) and planet carrier (7). Pry planetary ring gear (3) with dowels from rear of housing. Remove planetary front cover (2) and shim (1).

To disassemble planet carrier (7), first remove snap ring (5—Fig. 166), if used. Press planet pins (8—Fig. 165 or Fig. 166) forward out of carrier. Remove planet gears (10) with needle rollers (12) and thrust washers (9). Splined sleeve (15) is retained in hub of carrier (7) by retaining ring (16).

When reassembling, be sure to account for all of the needle rollers (12). Each pinion contains two paths of rollers separated by a washer (11). Each roller path contains 27 rollers (12—Fig. 165) on 365 models. Each roller path contains 16 rollers (12—Fig. 166) on 375, 390 and 398 models. Use petroleum jelly (not heavy grease) to hold rollers in place while pressing planet pins into carrier. Gap in snap ring (5) should be located between planet pins.

Concave side of Belleville washer (18—Fig. 166) should be toward rear. Be sure that front plate (2—

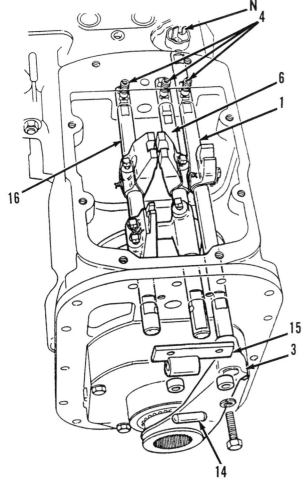

Fig. 163—Drawing of shift rails, installed. Neutral start switch (N) contacts flat on rail (1). Interlock plunger (14) is located in retainer (15). Refer to Fig. 164 for legend.

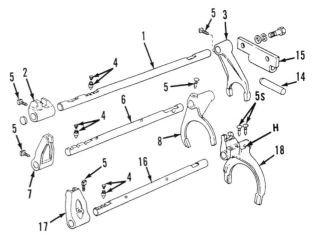

Fig. 164—Exploded view of shift rails and forks for twelve-speed multi-power transmission with synchronized second and third speeds in the main transmission.

1. Shift rail (High/Low)	
2. Shift gate	7. Shift gate
3. Shift fork	8. Shift fork
4. Detent springs	14. Interlock
and plungers	15. Retainer
5. Retaining screws	16. Shift rail (2nd/3rd)
6. Shift rail (1st	17. Shift gate
& Reverse)	18. Shift fork

Fig. 165 or Fig. 166) and rear plate (13) are positioned with grooved side toward planet carrier. Slots in front plate and shim (1) must be aligned and toward top when installed on transmission. Use petroleum jelly to hold thrust rings (6) in place. Make certain that

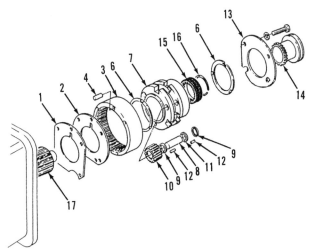

Fig. 165—Exploded view of high/low range planetary unit used on 365 model. Refer to Fig. 166 for other models with multi-power transmission.

1. Front shim	10. Pinion
2. Front plate	11. Spacer washer
3. Ring gear	12. Needle rollers (27/path)
4. Dowel	13. Rear plate
6. Thrust washer	14. Shift coupler
7. Planet carrier	15. Splined sleeve
8. Pinion shaft	16. Retaining ring
9. Side washer	17. Transmission mainshaft

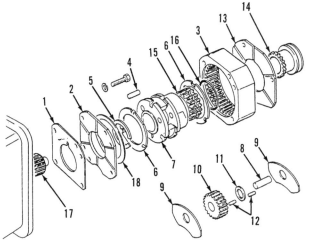

Fig. 166—Exploded view of high/low range planetary unit used on 375, 390 and 398 models. Refer to Fig. 165 for planetary unit used on other models with multi-power transmission.

1. Front shim	10. Pinion
2. Front plate	11. Spacer washer
3. Ring gear	12. Needle rollers (16/path)
4. Dowel	13. Rear plate
5. Snap ring	14. Shift coupler
6. Thrust washer	15. Splined sleeve
7. Planet carrier	16. Retaining ring
8. Pinion shaft	17. Transmission mainshaft
9. Side washer	18. Belleville washer

tangs of thrust rings engage notches in pinion carrier and that brass side faces away from carrier. On 365 models, cutaway section of rear plate (13—Fig. 165) should be positioned at the lower, **right** corner of planetary as shown. A lockwasher is used only on mounting screw at corner of cutaway. On 375, 390 and 398 models, the slanted section of rear plate (13—Fig. 166) should be positioned at lower, **left** corner of planetary. On all models, tighten mounting screws to 40-47 N•m (30-35 ft.-lbs.) torque. Complete assembly procedure by reversing disassembly.

PTO INPUT SHAFT AND RETAINER HOUSING

Models With Twelve-Speed Multi-Power Transmission

141. To remove the pto input shaft (8—Fig. 167) and retainer (3), first split tractor between engine and transmission as outlined in paragraph 93. Remove clutch release bearing (16), retainer (15), fork (13) and release shaft (12). Remove brake cross shaft, if so equipped. Disconnect pipe (3P) from fitting (3F). Remove screws and pull pto input shaft (8) and retainer (3) as a unit from transmission housing.

To disassemble, remove snap ring (6) from rear of retainer, then push input shaft and bearing (5) out of housing, toward rear. Bearing can be removed from input shaft after removing seal rings (8S) and snap ring (4). Remove oil seals (1 and 9) and needle bearing (2) only if new parts are to be installed.

Lubricate all seals and bearings with petroleum jelly before assembling. Use special tool (MF 315A or equivalent) to install new needle bearing (2). Install seal (9) with lip toward rear, using special tool (MF 256A or equivalent). Install ball bearing (5) on shaft with shield toward rear, then install snap ring (4). Install two new seal rings (8S) in grooves of shaft. Make sure that ends of seal rings are locked, coat shaft with clean transmission oil, then insert pto shaft into bore of retainer (3). Be sure that seal rings (8S) are not damaged while installing pto shaft, then install snap ring (6) and "O" ring (7). Use special tool (MF255B or equivalent) to install seal (1) with lip toward rear. Use special sleeve (KMF 1004 or equivalent) to protect seal (1), slide pto input shaft and retainer housing, over transmission input shaft (11). Coat threads of retaining screws with "Hylomar" or equivalent sealer and tighten to 54-61 N•m (40-45 ft.-lbs.) torque. Select thickness of thrust washer (10) that will limit clearance between thrust washer and pto input shaft gear to 0.63-1.65 mm (0.025-0.065 inch). Thrust washer (10) is available in thicknesses of 2.29-2.36, 2.92-3.00 and 3.43-3.50 mm (0.090-0.093, 0.115-0.118 and 0.035-0.138 inch). Reattach

Fig. 167—Pto input shaft (8) and retainer housing (3) can be removed as a unit from front of transmission housing. Thrust washer (10) is available in various thicknesses.

1. Seal
2. Needle bearing
3. Retainer housing
4. Snap ring
5. Ball bearing
6. Snap ring
7. "O" ring
8. Pto input shaft
8S. Seal rings
9. Oil seal
10. Thrust washer
11. Transmission input shaft
12. Clutch lever and shaft
13. Clutch release lever
14. Return spring
15. Release housing
16. Release bearing

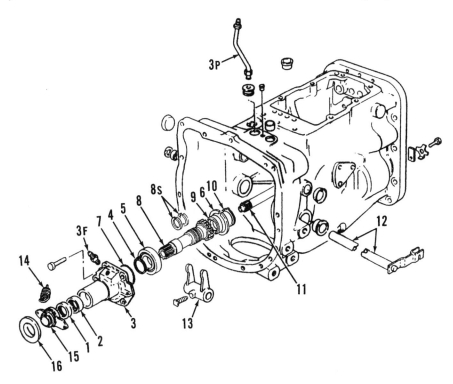

pipe (3P) to fitting (3F). Remainder of assembly is reverse of disassembly.

PTO LOWER SHAFT FRONT BEARING AND RETAINER

Models With Twelve-Speed Multi-Power Transmission

142. The pto lower shaft front bearing (41—Fig. 168) can be removed and installed after separating tractor between the rear of the engine and front of transmission housing. Removal of the lower driven gear (45) and shaft (46) require the additional steps of separating at rear of transmission housing and removal of the transmission top cover and shift rails. The following procedure describes only service to the front bearing and bearing retainer (42) after splitting as outlined in paragraph 93.

Remove clutch release bearing (16—Fig. 167), retainer (15), fork (13) and release shaft (12). Unbolt and remove cover (36—Fig. 168), then remove snap ring (38) and washer (39). Thread a ⅜ inch UNC cap screw, 75 mm (3 inches) long into each of the two threaded holes of retainer (42). Tighten the two jack-

Fig. 168—Exploded view of the pto lower shaft and bearings.

36. Cover
37. "O" ring
38. Snap ring
39. Washer
40. Snap ring
41. Ball bearing
42. Retainer
43. Gasket
44. Snap ring
45. Pto lower gear
46. Pto lower shaft

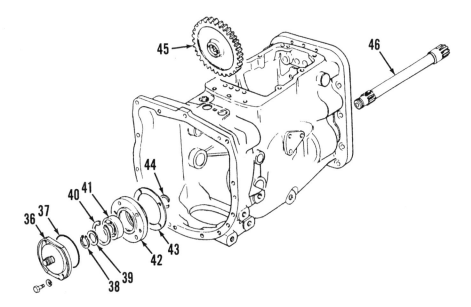

121

screws to push retainer (42) and bearing (41) forward off shaft (46). Bearing can be pressed from retainer after removing snap ring (40). Remove "O" ring (37) and gasket (43), then clean all parts thoroughly.

Coat gasket (43) and threads of retaining screws lightly with "Hylomar" or equivalent sealer before installing. End of shaft (46) is threaded to accept a puller screw to pull bearing (41) onto shaft until washer (39) and snap ring (38) can be installed. Align holes in retainer (42) with holes in transmission housing, install "O" ring (37) and cover (36). Tighten retaining screws to 54-61 N·m (40-45 ft.-lbs.) torque. Remainder of assembly is reverse of disassembly procedure.

TRANSMISSION INPUT SHAFT AND MULTI-POWER CLUTCH

Models With Twelve-Speed Multi-Power Transmission

143. To remove the transmission input shaft (11—Fig. 167), first remove the transmission as outlined in paragraph 137. Remove the top cover and shift levers as outlined in paragraph 138, planetary unit

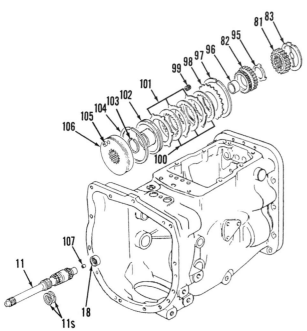

Fig. 169—Exploded view of transmission input shaft and multi-power clutch.

11. Transmission input shaft	
11S. Input shaft seal rings	99. Release springs (6 used)
18. Needle bearing	100. Internal discs (3 used)
81. Main drive gear	101. External plates (3 used)
82. Overdrive gear and hub	102. Clutch piston
83. Spacer	103. Seal ring
95. Thrust washer	104. Seal ring
96. Bushing	105. Check valve
97. Snap ring	106. Clutch drum
98. Plate	107. Plug

as outlined in paragraph 140 and the shift rails and forks as outlined in paragraph 139. Refer to paragraph 141 and remove the pto input shaft and retainer. Slide input shaft (11—Fig. 169) forward, then withdraw gears (81 and 82), thrust washers (83 and 95) and multi-power clutch (97 through 106) out top opening. Refer to paragraph 141 for service to the pto input shaft and housing.

To disassemble the multi-power clutch, press plate (98) into clutch and remove snap ring (97). Release plate (98) and withdraw clutch plates (98 and 101) and discs (100). Carefully remove piston (102) from bore of clutch drum. Install new seal rings (103 and 104) if removed.

Friction discs (100) should be 2.41-2.59 mm (0.095-0.102 inch) thick and groove depth should be 0.38-0.63 mm (0.015-0.025 inch). Measure distortion by measuring maximum height which should not exceed 2.92 mm (0.115 inch).

Steel plates (101) should be 1.67-1.75 mm (0.66-0.69 inch) thick, should have maximum dish of 0.25 mm (0.010 inch) and should not be scored or overheated. Measure distortion by measuring maximum height which should not exceed 2.21 mm (0.0875 inch).

Springs (99) should have free length of 17.8 mm (0.70 inch). Springs should exert 2.98-3.64 kg (6.75-8.03 lbs.) when compressed to normal working height of 12.7 mm (0.50 inch).

Reassemble multi-power clutch piston (102) and seal rings (103 and 104) in drum (105). Install one of the steel plates (101) with the external lugs in the groove just to the right of the six holes in drum (106). Position one of the six springs (99) on each of the lugs of the first plate installed. Install one of the internally splined friction discs (100), followed by the second steel plate (101). The second steel plate should have lugs in the groove just to the right of the lug for the first plate. Install the second internally splined friction disc (100), followed by the third steel plate (101). The third steel plate should have lugs in the groove just to the right of the lugs for the second plate. Install the third internally splined friction disc (100), followed by retainer plate (98). Compress springs with retainer plate and install snap ring (97).

Align internal splines of friction discs (100) so that gear hub (82) can be inserted. Use petroleum jelly to stick thrust washer (95) to rear of gear (82). Make sure that tabs of thrust washer correctly engage lugs of gear. Install new seal rings (11S) in grooves of input shaft (11), making sure that ends are properly hooked together. Position multi-power clutch assembly, gears (81 and 82) and thrust washers (83 and 95) in position and insert shaft (11). Remainder of installation is the reverse of removal procedures. Thrust washer (10—Fig. 167) is available in three thicknesses to adjust shaft end play. Refer to paragraph 141 for measuring and adjusting shaft end play. Refer to paragraph 139 for adjusting neutral position of forks on shift shafts.

MAINSHAFT (OUTPUT SHAFT) AND GEARS

Models With Twelve-Speed Multi-Power Transmission

144. To remove the mainshaft (19—Fig. 170 and Fig. 171), first remove the transmission as outlined in paragraph 137, the transmission top cover as outlined in paragraph 138, the planetary reduction unit as outlined in paragraph 140, the shift rails and forks as outlined in paragraph 139, the pto input shaft as outlined in paragraph 141 and the multi-power clutch and transmission input shaft as outlined in paragraph 143.

Remove snap ring (14—Fig. 171) and slide coupler (91) forward, engaging 1st gear (13). Withdraw mainshaft (19) toward rear and bearing (15) and thrust washer (17) toward front. Lift 1st gear (13), bushing (90), coupling (89), ring (75), hub (92) and sliding coupling (91) out top opening, then remaining parts of mainshaft can be removed.

Synchronizer parts (64 through 67) should be removed as an assembly and separated only if necessary. Wrap synchronizer assembly with a cloth to catch poppet parts (65 and 66) when coupling (64) is removed from hub (67). Clean, inspect and renew any components that are damaged or worn excessively. Place synchronizer cones (72 and 73) and synchronizer rings (63 and 68) on respective gears (21 and 24). Make sure that each ring is seated squarely on taper of cone. Measure clearance between ring and gear at several places around gear, using a feeler gauge. Install new synchronizer ring (63 or 68) if clearance is less than 0.5 mm (0.020 inch). This clearance can be checked with gearbox assembled, but minimum clearance is increased to 0.8 mm (0.030 inch).

When reassembling the synchronizer shift components, align centralizing holes in shift collar (64) and hub (67). Special tools (MF 415 and MS 550) are available to assist holding the three poppets (65 and 66) into hub (67) while installing coupling (64).

Assemble mainshaft (19), snap rings (14 and 23), bearings (15 and 22) and all parts between snap rings (14 and 23), outside of gearbox housing. Refer to Fig. 172 for cross section of assembled mainshaft. Measure clearance (G) between front snap ring (14) and inner race of bearing (15). Select spacer (17—Fig. 171 and Fig. 172) of proper thickness to limit gap (G) to 0.080-0.300 mm (0.003-0.012 inch). Spacer (17) is available in four thicknesses of 4.14-4.19, 4.39-4.44, 4.62-4.67 and 4.85-4.90 mm (0.163-0.165, 0.173-0.175, 0.182-0.184 and 0.191-0.193 inch). Reassemble

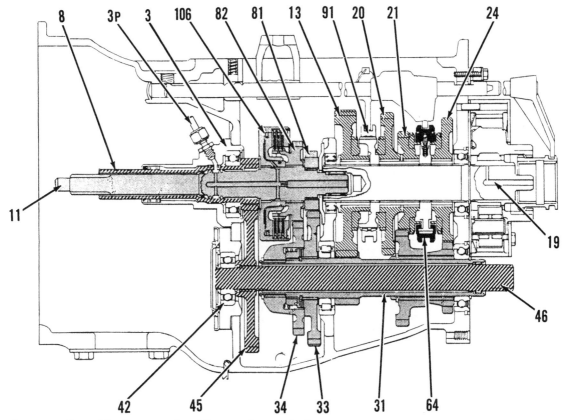

Fig. 170—Cross section of twelve-speed multi-power transmission.

mainshaft and components inside housing using the preselected spacer (17). Shorter sleeve (70—Fig. 171) is in 2nd gear (24) and longer sleeve (61) is in 3rd and reverse gears (20 and 72). Flat side of spacer (17) should be toward gear (13).

Remainder of assembly is reverse of disassembly procedure. Coat threads of screws attaching retainer housing (3—Fig. 167) lightly with "Hylomar" or equivalent sealer before tightening to 54-61 N•m (40-45 ft.-lbs.) torque.

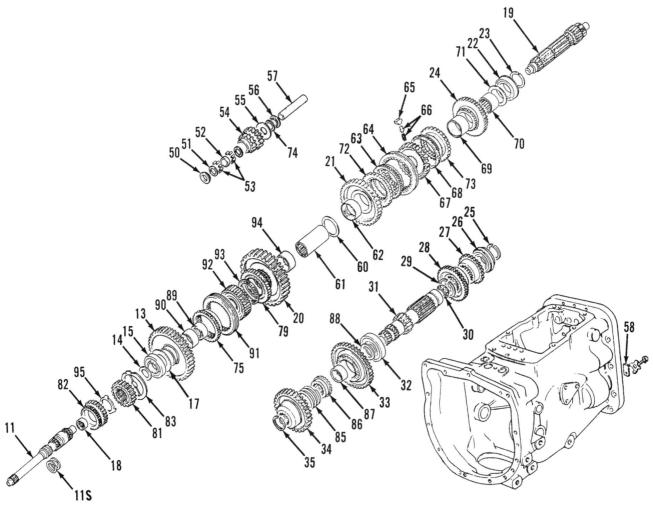

Fig. 171—Exploded view of twelve-speed multi-power transmission.

11. Transmission input shaft	29. Snap ring	61. Sleeve (longer)	74. Spring
11S. Seal rings	30. Roller bearing	62. Bushing	75. Shifter ring
13. Gear (1st)	31. Countershaft &	63. Synchronizer ring	79. Shifter hub
14. Snap ring	1st gear	(Same as 68)	81. Gear (Low/Tortoise)
15. Bearing	32. Bearing	64. Shifting coupler	82. Gear (High/Hare)
17. Spacer (variable	33. Gear (Low/Tortoise)	65. Pressure blocks (3 used)	83. Spacer
thicknesses)	34. Gear (High/Hare)	66. Ball pins & springs (3 used)	85. Spring
18. Roller bearing	35. Snap ring	67. Synchronizer hub	86. Coupler
19. Mainshaft	50. Thrust washer	68. Synchronizer ring	87. Bushing
20. Gear (Reverse)	51. Washers	(Same as 63)	88. Thrust washer
21. Gear (3rd)	52. Center spacer	69. Bushing	89. Sleeve (medium)
22. Ball bearing	53. Needle rollers	70. Sleeve (short)	90. Bushing
23. Snap ring	54. Gear (Reverse idler)	71. Washer	91. Shifting coupler
24. Gear (2nd)	55. Friction disc	72. Synchronizer cone	92. Hub
25. Snap ring	56. Spacer	(Same as 73)	93. Thrust bearing
26. Ball bearing	57. Idler shaft	73. Synchronizer cone	94. Bushing
27. Gear (2nd)	58. Retaining clip	(Same as 72)	95. Thrust bearing
28. Gear (3rd)	60. Washer		

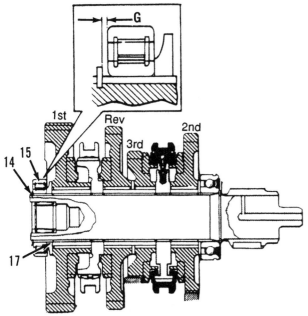

Fig. 172—The mainshaft should be assembled outside of transmission housing to determine correct thickness of thrust washer (17) as determined by gap (G).

COUNTERSHAFT (LAYSHAFT) AND GEARS

Models With Twelve-Speed Multi-Power Transmission

145. The mainshaft must be removed as outlined in paragraph 144 and pto lower shaft should be removed as outlined in paragraph 142, before removing countershaft. Remove snap ring (25—Fig. 171) from rear, bump shaft forward enough to unseat snap ring (29) from groove and move snap ring forward to unsplined part of countershaft. Remove snap ring (35), then tap shaft rearward. Remove overdrive gear (34), spring (85), jaw coupler (86), drive gear (33) and thrust washer (88) out top opening. Bump shaft forward and lift gears (27 and 28) out.

Reassemble by reversing removal procedure. Always install new snap rings when reassembling. Position snap ring (29) around shaft in the unsplined area and install front bearing (32) on shaft before installing shaft in housing. Insert countershaft from the front and position the two gears (27 and 28) on rear of shaft with their hubs facing together. Move shaft to rear and install thrust washer (88), drive gear (33), jaw coupling (86), spring (85), overdrive gear (34) and snap ring (35). Move shaft forward and push 2nd and 3rd gears (27 and 28) to rear so that center snap ring (29) can be seated in groove. Push shaft to rear and block in place, then install rear bearing (26) and snap ring (25).

Refer to paragraph 144 for assembly of mainshaft and selection of thrust washer (17—Fig. 170 and Fig. 171). Refer to Fig. 170 for cross section of assembled transmission. Coat gasket (43—Fig. 168) lightly with "Hylomar" or equivalent sealer before installing. Coat threads of screws attaching cover (36) and screws attaching retainer housing (3—Fig. 167) lightly with "Hylomar" or equivalent sealer before tightening to 54-61 N·m (40-45 ft.-lbs.) torque.

REVERSE IDLER GEAR AND SHAFT

Models With Twelve-Speed Multi-Power Transmission

146. A dummy shaft should be fabricated to hold parts in place while both removing and installing the reverse idler shaft. Dummy shaft should be 25 mm (1 inch) diameter and 55 mm (2 3/16 inches) long.

To remove the reverse idler, first remove the mainshaft and countershaft assemblies as outlined in paragraphs 144 and 145. Release locking washer, then remove locking screw and clip (58—Fig. 171). Slide dummy shaft in from front, pushing reverse idler shaft (57) out toward rear. Dummy shaft should be used to hold needle rollers and washers in place while both removing and installing the reverse idler. Be sure to remove all 56 of the needle rollers.

Assemble needle rollers, washers and spacer in reverse idler using petroleum jelly and the dummy shaft. Assemble center spacer (52), one path of 28 needle rollers (53), one end washer (51) and one thrust washer (50), around dummy shaft, in one end of reverse idler gear. Assemble the second path of 28 needle rollers, the other end washer (51) and the remaining thrust washer (50). Center the dummy shaft so that it holds both thrust washers in place, then position assembly in housing and install idler shaft (57). Install clip (58), lock plate and retaining screw. Tighten retaining screw and lock by bending tab of lock plate against flat of screw. Complete reassembly by reversing disassembly procedure.

TRANSMISSION
(EIGHT-SPEED SHUTTLE)

362, 365 and 375 Models

Eight forward speeds are possible by using a standard gearbox with four forward speeds compounded by a two speed planetary unit. An equal number of reverse speeds is made possible by fitting a forward/reverse unit in front of the gearbox. Refer to Fig. 173 for shift pattern of eight-speed shuttle transmission available in 362, 365 and 375 model tractors.

LUBRICATION

Models With Eight-Speed Shuttle Transmission

147. The transmission, rear differential and hydraulic system share common fluid which is contained in the transmission and center housing. Recommended lubricant is Massey-Ferguson Super 500 multi-use 10W-30 oil, Massey-Ferguson Permatran oil or equivalent transmission/hydraulic oil. Capacity is 47.4 L (12.5 gal.) if equipped with spacer or four-wheel-drive transfer box between transmission and rear axle center housing; 43.4 L (11.5 gal.) if not equipped with transfer box or spacer. Tractors are equipped with dipstick (Fig. 174 or Fig. 175) with marks indicating maximum and minimum oil levels. Level should be maintained at maximum level if tractor is operated on hilly terrain or if using implements which require large quantities of oil.

TRACTOR REAR SPLIT

Models With Eight-Speed Shuttle Transmission

148. To separate the rear of the transmission from the rear axle center housing (Fig. 176), refer to paragraph 99 if tractor does not have cab or to paragraph 100 if equipped with cab.

TRANSMISSION REMOVAL

Models With Eight-Speed Shuttle Transmission

149. To remove the transmission, first separate the rear of the transmission from rear axle center hous-

ing as outlined in paragraph 99 if tractor does not have cab or to paragraph 100 if equipped with cab. If not equipped with cab, unbolt and support or remove the steering and instrument console. On all models, support the engine and transmission housing separately. Remove screws and stud nuts, then separate transmission from engine.

Reassemble transmission to engine by reversing the splitting procedure and observe the following. Install a M12 alignment stud approximately 100 mm

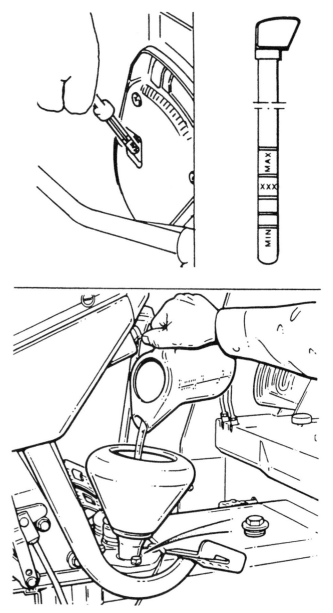

Fig. 174—Transmission oil level of tractors without cab is measured by dipstick located in the side cover as shown. Refer to Fig. 175 for models with cab.

Fig. 173—Shift pattern for eight-speed shuttle transmission.

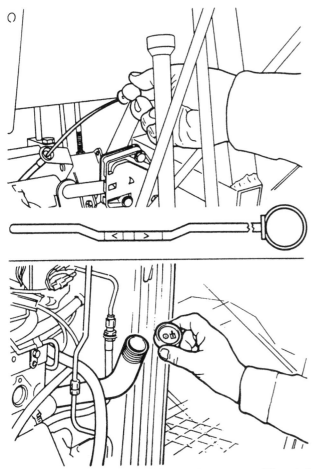

Fig. 175—Transmission oil level of tractors with cab is measured by dipstick that points toward rear as shown.

(4 inches) long in each side of the transmission housing to assist in aligning engine to transmission. Turn flywheel to align clutch plate splines with transmission and pto input shaft splines. Install retaining screws and nuts after engine and transmission housing flanges are completely together. Tighten retaining screws and nuts to 115 N•m (85 ft.-lbs.) torque.

Refer to paragraph 99 or 100 and reconnect transmission to rear axle center housing. Tighten all screws retaining the transmission housing to spacer housing, transfer case or rear axle center housing to 112 N•m (83 ft.-lbs.) torque, beginning at top center and progressing in a clockwise direction (as viewed from tractor rear) two times around the mating flange. On four-wheel-drive models, coat threads of drive shaft retaining screws with "Loctite 270," then attach drive shaft flanges and tighten screws to 55-75 N•m (40-55 ft.-lbs.) torque. On all models, check and adjust clutch linkage as outlined in paragraph 92.

TOP COVER

Models With Eight-Speed Shuttle Transmission

150. The shift levers and the small shift tower can be removed after first removing floor mat and access panels. Disconnect forward and reverse shift linkage from shift lever. Installation of the shift levers and tower is easier if transmission is shifted into neutral before removing. Shift tower (11—Fig. 177) is attached to the top cover (18) by five screws.

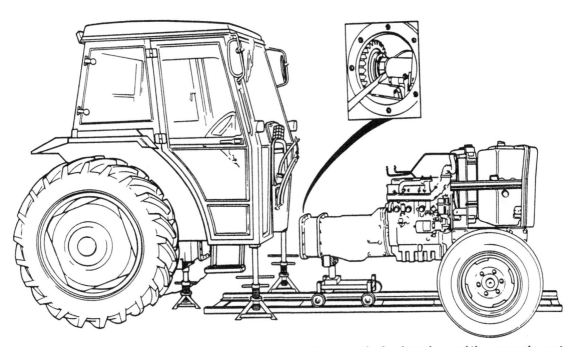

Fig. 176—View of tractor with standard cab separated between the transmission housing and the rear axle center housing. Special splitting stand and track is shown. Support and splitting procedures are similar for other models.

The main shift lever (6) can be removed after removing the tower assembly, cover (3), snap ring (9), spring (8) and pin (10). The high/low shift lever (13) and forward/reverse shift lever (19) can be removed after removing cover (14) and pin (16).

Refer to Fig. 178 for forward/reverse shuttle shift linkage. Be sure that electrical wiring does not touch any part of shift linkage. Lubricate linkage with general purpose grease when assembling. Adjust lengths of rods attached to the relay lever as shown in Fig. 179. Install shims between relay lever pivot and retaining snap ring to remove any end play that would result in lost motion.

The transmission top cover (18—Fig. 177) can be unbolted and lifted from transmission after first re-

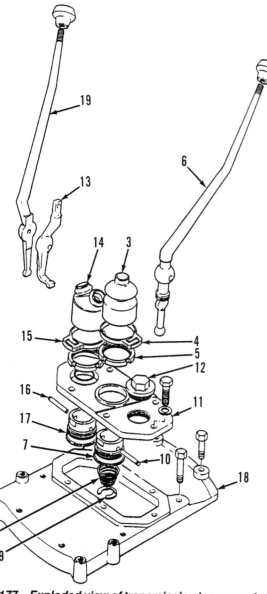

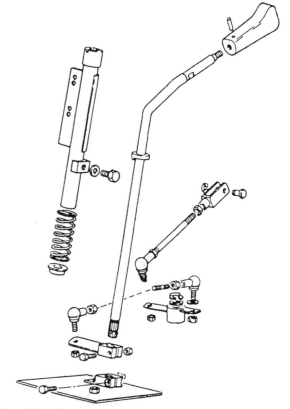

Fig. 178—Forward/Reverse shift lever (13—Fig. 177) is operated by linkage similar to type shown.

Fig. 177—Exploded view of transmission top cover, tower cover and shift levers typical of models with eight-speed shuttle transmission.

3. Cover	12. Fill plug
4. Retainer clip	13. Forward/Reverse
5. Retaining nut	shift lever
6. Main shift lever	14. Cover
7. Sleeve	15. Retaining nut
8. Spring	16. Pin
9. Snap ring	17. Sleeve
10. Pin	18. Top cover
11. Tower cover	19. High/Low shift lever

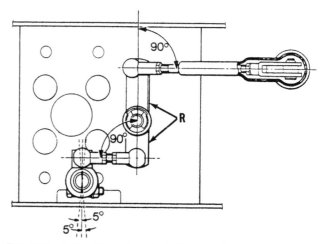

Fig. 179—Adjust lengths of shift control rods attached to the relay lever (R) to provide alignment of forward/reverse shift linkage as shown.

moving the shift levers and tower assembly. On models with cab, it is necessary to split rear of transmission from rear axle center housing as outlined in paragraph 100 or remove the cab assembly. Models without cab must have instrument console and other interfering parts removed before transmission top cover. The reason for removing top cover will determine what additional procedures are necessary.

"Loctite 515 Instant Gasket" or equivalent should be used to seal tower to top cover. Tighten retaining screws to 50-70 N•m (37-52 ft.-lbs.) torque.

SHIFTER RAILS AND FORKS

Models With Eight-Speed Shuttle Transmission

151. To remove the shifter rails and forks, first refer to paragraph 99 or paragraph 100 and separate tractor between rear of transmission and front of axle center housing. Refer to paragraph 150 and remove the shift lever and tower assembly and transmission top cover. Models without cab must have instrument

and steering console removed before top cover can be unbolted and removed.

On all models, remove the neutral safety start switch (N—Fig. 181). Remove safety wire and set screws (5—Fig. 180) and lift detent springs and plungers (4) from bores in housing. Unbolt and remove interlock parts (12, 13, 14 and 15—Fig. 181). Slide rail out of housing toward rear and remove gate and shift fork through top. Twist rails if necessary while withdrawing. Fork (8—Fig. 180) can not be removed until transmission input shaft and gears are removed as outlined in paragraph 155.

When assembling, lubricate rails and make sure that rails do not have burrs, especially around the set screw locations, that would prevent easy and smooth installation. Each of the four shift rails is different from the other rails. Tighten screws retaining the interlock retainer (15—Fig. 181) to 40-47 N•m (30-35 ft.-lbs.) torque and the seven set screws (5) to 34-52 N•m (25-38 ft.-lbs.) torque. Install safety wire through set screws (5) after tightening to prevent loosening. Adjust position of shift fork (18) as follows.

Use special tools (MF 414/2 and MF 414/3 or equivalent) to hold shift rails (6, 9 and 16—Fig. 180

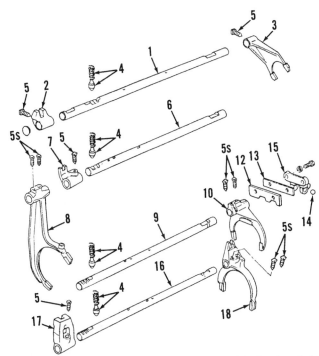

Fig. 180—Exploded view of shift rails and forks for eight-speed shuttle transmission.

1. Shift rail (High/Low)	9. Shift rail (1st/2nd)
2. Shift gate	10. Shift fork
3. Shift fork	12. Plate
4. Detent springs and plungers	13. Gasket
5. Retaining screws	14. Interlock plunger & balls
6. Shift rail (Forward/Reverse)	15. Retainer
7. Shift gate	16. Shift rail (3rd/4th)
8. Shift fork	17. Shift gate
	18. Shift fork

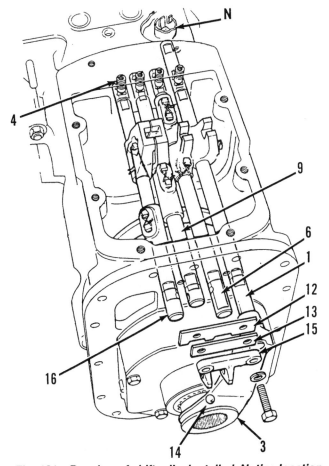

Fig. 181—Drawing of shift rails, installed. Notice location of neutral start switch (N). Interlock balls (14) are located in retainer (15). Refer to Fig. 180 for legend.

and Fig. 181) in neutral detent position. Insert centralizing tool (MF 414/1 or equivalent) into holes in shift forks (8, 10 and 18—Fig. 180) and into matching hole of sliding synchronizing coupling. Tighten each set screws (5S), while turning centralizing tool, to hold the fork in this center position. Screws (5 and 5S) should be tightened to 34-52 N•m (25-38 ft.-lbs.) torque and locked in position with safety wire. Shift all gears, rails and gates to neutral after checking to make sure that all shift gates align. Shift into each gear to make sure that each can be correctly engaged. Install detent plungers and springs (4—Fig. 180). Remainder of assembly is the reverse of disassembly procedure.

PLANETARY UNIT

Models With Eight-Speed Shuttle Transmission

152. To remove the planetary unit, first refer to paragraph 99 or paragraph 100 and separate tractor between rear of transmission and front of axle center housing. Remove set screw (5—Fig. 180) from fork (3), then remove the fork and coupler from rear of planetary. Remove the four retaining cap screws and withdraw rear cover (13—Fig. 182 or Fig. 183), rear thrust ring (6) and planet carrier (7). Pry planetary ring gear (3) with dowels from rear of housing. Remove planetary front cover (2) and shim (1).

To disassemble planet carrier (7), first remove snap ring (5—Fig. 183), if used. Press planet pins (8—Fig. 182 or Fig. 183) forward out of carrier. Remove planet gears (10) with needle rollers (12) and thrust washers (9). Splined sleeve (15) is retained in hub of carrier (7) by retaining ring (16).

When reassembling, be sure to account for all of the needle rollers (12). Each pinion contains two paths of rollers separated by a washer (11). Each roller path contains 27 rollers (12—Fig. 182) on 362 and 365 models. Each roller path contains 16 rollers (12—Fig. 183) on 375 and 390 models. Use petroleum jelly (not heavy grease) to hold rollers in place while pressing planet pins into carrier. Gap in snap ring (5—Fig. 183) should be located between planet pins.

Concave side of Belleville washer (18—Fig. 183) should be toward rear. Be sure that front plate (2—Fig. 182 or Fig. 183) and rear plate (13) are positioned with grooved side toward planet carrier. Slots in front plate and shim (1) must be aligned and toward top when installed on transmission. Use petroleum jelly to hold thrust rings (6) in place. Make certain that tangs of thrust rings engage notches in pinion carrier and that brass side faces away from carrier. On 362 and 365 models, cutaway section of rear plate (13—Fig. 182) should be positioned at the lower, **right** corner of planetary as shown. A lockwasher is used only on mounting screw at corner of cutaway. On 375 and 390 models, the slanted section of rear plate (13—Fig. 183) should be positioned at lower, **left**

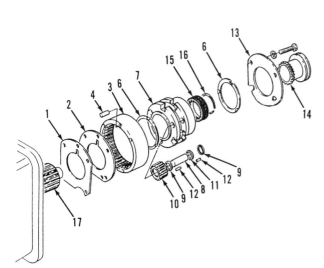

Fig. 182—Exploded view of high/low range planetary unit used on 362 and 365 models. Refer to Fig. 183 for other models.

1. Front shim
2. Front plate
3. Ring gear
4. Dowel
6. Thrust washer
7. Planet carrier
8. Pinion shaft
9. Side washer
10. Pinion
11. Spacer washer
12. Needle rollers (27/path)
13. Rear plate
14. Shift coupler
15. Splined sleeve
16. Retaining ring
17. Transmission mainshaft

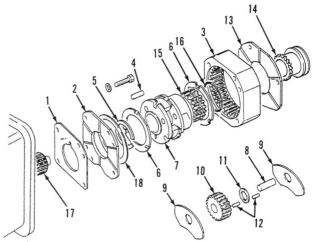

Fig. 183—Exploded view of high/low range planetary unit used on 375 and 390 models. Refer to Fig. 182 for planetary unit used on other models with eight-speed shuttle transmission.

1. Front shim
2. Front plate
3. Ring gear
4. Dowel
5. Snap ring
6. Thrust washer
7. Planet carrier
8. Pinion shaft
9. Side washer
10. Pinion
11. Spacer washer
12. Needle rollers (16/path)
13. Rear plate
14. Shift coupler
15. Splined sleeve
16. Retaining ring
17. Transmission mainshaft
18. Belleville washer

corner of planetary. On all models, tighten mounting screws to 40-47 N·m (30-35 ft.-lbs.) torque. Complete assembly procedure by reversing disassembly.

PTO INPUT SHAFT AND RETAINER HOUSING

Models With Eight-Speed Shuttle Transmission

153. To remove the pto input shaft (8—Fig. 184) and retainer (3), first split tractor between engine and transmission as outlined in paragraph 93. Remove clutch release bearing (16), retainer (15), fork (13) and release shaft (12). Remove brake cross shaft, if so equipped. Remove screws and pull pto input shaft (8) and retainer (3) as a unit from transmission housing.

To disassemble, remove snap ring (6) from rear of retainer, then push input shaft and bearing (5) out of housing toward rear. Bearing can be removed from input shaft after removing snap ring (4). Remove oil seals (1 and 9) and needle bearing (2) only if new parts are to be installed.

Lubricate all seals and bearings with petroleum jelly before assembling. Use special tool (MF 315A or equivalent) to install new needle bearing (2). Install seal (9) with lip toward rear, using special tool (MF 421 or equivalent). Install ball bearing (5) on shaft with shield toward rear, then install snap ring (4). Use special tool (MF255B or equivalent) to install seal (1) with lip toward rear. Use special sleeve (KMF 1004 or equivalent) to protect seal (1), slide input shaft into the retainer housing, then install snap ring (6). Install "O" ring (7) and install housing and shaft assembly over transmission input shaft (11) into housing bore. Coat threads of retaining screws with

"Hylomar" or equivalent sealer and tighten to 54-61 N·m (40-45 ft.-lbs.) torque. Remainder of assembly is reverse of disassembly.

PTO LOWER SHAFT FRONT BEARING AND RETAINER

Models With Eight-Speed Shuttle Transmission

154. The pto lower shaft front bearing (41—Fig. 185) can be removed and installed after separating tractor between the rear of the engine and front of transmission housing. Removal of the lower driven gear (45) and shaft (46) require the additional steps of separating at rear of transmission housing and removal of the transmission top cover and shift rails. The following procedure describes only service to the front bearing and bearing retainer (42) after splitting as outlined in paragraph 93.

Remove clutch release bearing (16—Fig. 184), retainer (15), fork (13) and release shaft (12). Unbolt and remove cover (36—Fig. 185), then remove snap ring (38) and washer (39). Thread a 3/8 inch UNC cap screw, 75 mm (3 inches) long into each of the two threaded holes of retainer (42). Tighten the two jack-screws to push retainer (42) and bearing (41) forward off shaft (46). Bearing can be pressed from retainer after removing snap ring (40). Remove "O" ring (37) and gasket (43), then clean all parts, thoroughly.

Coat gasket (43) and threads of retaining screws lightly with "Hylomar" or equivalent sealer before installing. End of shaft (46) is threaded to accept a puller screw to pull bearing (41) onto shaft until washer (39) and snap ring (38) can be installed. Align holes in retainer (42) with holes in transmission housing, install "O" ring (37) and cover (36). Tighten

Fig. 184—Pto input shaft (8) and retainer housing (3) can be removed as a unit from front of transmission housing.

1. Seal
2. Needle bearing
3. Retainer housing
4. Snap ring
5. Ball bearing
6. Snap ring
7. "O" ring
8. Pto input shaft
9. Oil seal
10. Thrust washer
11. Transmission input shaft
12. Clutch lever and shaft
13. Clutch release lever
14. Return spring
15. Release housing
16. Release bearing

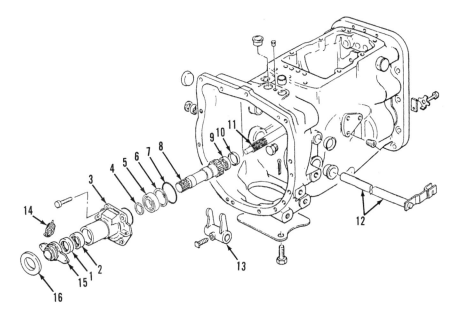

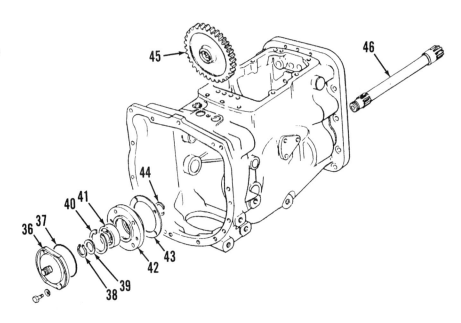

Fig. 185—Exploded view of the pto lower shaft and bearings.

36. Cover
37. "O" ring
38. Snap ring
39. Washer
40. Snap ring
41. Ball bearing
42. Retainer
43. Gasket
44. Snap ring
45. Pto lower gear
46. Pto lower shaft

retaining screws to 54-61 N•m (40-45 ft.-lbs.) torque. Remainder of assembly is reverse of disassembly procedure.

TRANSMISSION INPUT SHAFT

Models With Eight-Speed Shuttle Transmission

155. To remove the transmission input shaft (11—Fig. 186 and Fig. 187), first remove the transmission as outlined in paragraph 149. Remove the top cover and shift levers as outlined in paragraph 150, planetary unit as outlined in paragraph 152 and the shift rails and forks as outlined in paragraph 151. Refer to paragraph 153 and remove the pto input shaft and retainer. Slide input shaft (11—Fig. 187) forward, then withdraw gears (81 and 82) out top. Forward/reverse shift fork (8—Fig. 180) can be removed after removing input gears.

Reinstall input shaft by reversing removal procedures. Refer to paragraph 151 for adjusting neutral position of forks on shift shafts.

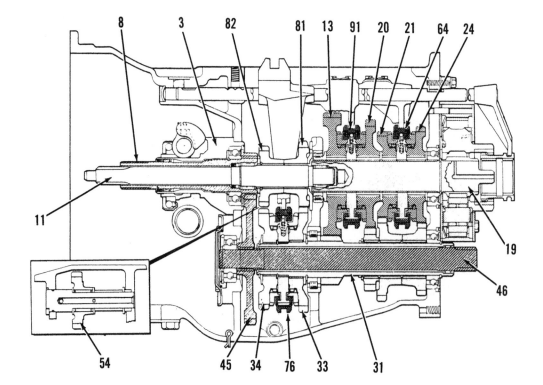

Fig. 186—Cross section of eight-speed shuttle transmission.

MAINSHAFT (OUTPUT SHAFT) AND GEARS

Models With Eight-Speed Shuttle Transmission

156. To remove the mainshaft (19—Fig. 186 and Fig. 187), first remove the transmission as outlined in paragraph 149. Remove the top cover and shift levers as outlined in paragraph 150, planetary unit as outlined in paragraph 152 and the shift rails and forks as outlined in paragraph 151. Refer to paragraph 153 and remove the pto input shaft and retainer, then remove the transmission input shaft as outlined in paragraph 155.

Remove spacer (83—Fig. 187), lift snap ring (14) from groove, push mainshaft (19) to rear and remove

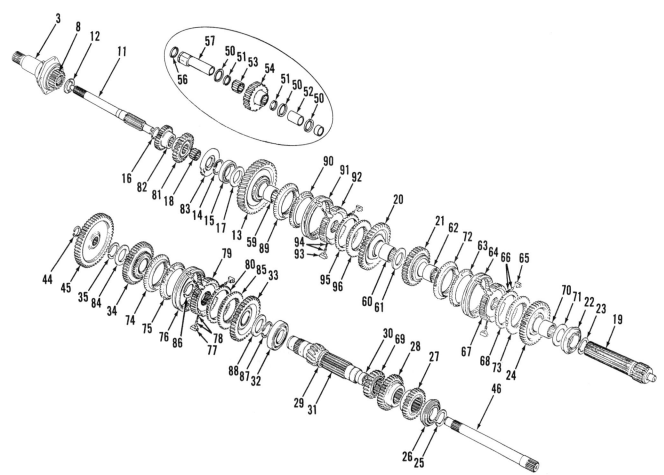

Fig. 187—Exploded view of eight-speed shuttle transmission. Synchronizer (63 through 68) is made up of same components as synchronizer (75 through 80).

3. Retainer	27. Gear (3rd)	60. Sleeve	77. Pressure blocks (3 used)
8. Pto input shaft and gear	28. Gear (4th)	61. Washer	78. Ball pins & springs (3 used)
11. Transmission input shaft	29. Snap ring	62. Sleeve	79. Synchronizer hub
	30. Roller bearing	63. Synchronizer ring (Same as 68)	80. Synchronizer ring (Same as 75)
12. Spacer	31. Countershaft & 1st gear	64. Shifting coupler	81. Forward gear
13. Gear (1st)	32. Ball bearing	65. Pressure blocks (3 used)	82. Reverse gear
14. Snap ring	33. Forward gear	66. Ball pins & springs (3 used)	83. Spacer
15. Bearing	34. Reverse gear	67. Synchronizer hub	84. Thrust washer
16. Snap ring	35. Snap ring	68. Synchronizer ring (Same as 63)	85. Synchronizer cone
17. Spacer (Variable thicknesses)	44. Snap ring	69. Gear (2nd)	86. Snap ring
18. Roller bearing	45. Pto driven gear	70. Sleeve	87. Snap ring
19. Main shaft	46. Pto lower shaft	71. Washer	88. Thrust washer
20. Gear (2nd)	50. Thrust washers	72. Synchronizer cone	89. Synchronizer cone
21. Gear (4th)	51. Washers	73. Synchronizer cone	90. Synchronizer ring
22. Ball bearing	52. Spacer	74. Synchronizer cone	91. Shifting coupler
23. Snap ring	53. Needle rollers	75. Synchronizer ring (Same as 80)	92. Hub
24. Gear (3rd)	54. Gear (Reverse idler)	76. Shifting coupler	93. Pressure blocks (3 used)
25. Snap ring	56. "O" ring		94. Ball pins & springs (3 used)
26. Ball bearing	57. Idler shaft		95. Synchronizer ring
	59. Sleeve		96. Synchronizer cone

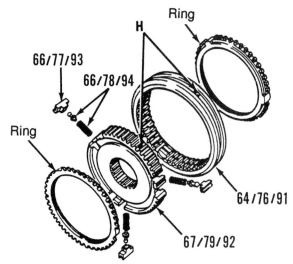

Fig. 188—Synchronizer rings for unit used on counter-shaft are different than similar parts used on mainshaft. Holes (H) are used when checking and adjusting position of shift forks and must be aligned when assembling synchronizer. Refer to Fig. 187 for legend.

gear (13). Continue to move mainshaft toward rear while lifting synchronizer assembly (90 through 95), thrust washer (61), gears (20, 21 and 24) and synchronizer assembly (63 through 68) out top opening. Rear bearing (22) can be removed if necessary after removing snap ring (23).

Synchronizer parts (64 through 67 and 90 through 95) should be removed as an assembly and separated only if necessary. Wrap synchronizer assembly with a cloth to catch poppet parts (65 and 66 or 77 and 78) when sliding coupling (64 or 76) is removed from hub (67 or 79). Clean, inspect and renew any components that are damaged or worn excessively. Place synchronizer cones (72 and 73 or 89 and 96) and synchronizer rings (63 and 68 or 90 and 95) on respective gears (21 and 24 or 13 and 20). Make sure that each ring is seated squarely on taper of cone. Measure clearance between ring and gear at several places around gear, using a feeler gauge. Install new synchronizer ring (63, 68, 90 or 95) if clearance is less than 0.5 mm (0.020 inch). This clearance can be checked with gearbox assembled, but minimum clearance is then increased to 0.8 mm (0.030 inch).

When reassembling the synchronizer shift components, align centralizing holes (H—Fig. 188) in shift collar (64 or 91) and hub (67 or 92). Special tools (MF 415 and MS 550) are available to assist holding the three poppets (65 or 93) into hub (67 or 92), against spring pressure, while installing coupling (64 or 91). Synchronizer rings (75 and 80) on countershaft are different than similar parts on mainshaft, because countershaft rotates in opposite direction.

Make sure that forward and reverse shift fork (8—Fig. 180) is in place in shift collar before installing the mainshaft.

Reassemble mainshaft (19—Fig. 187), snap rings (14 and 23), bearings (15 and 22) and all parts between snap rings (14 and 23), outside of gearbox housing. Be sure that sleeves (59, 60, 62 and 70) match respective gears (13, 20, 21 and 24) in length. Refer to Fig. 186 for cross section of assembled transmission. Measure clearance between front snap ring (14—Fig. 187) and inner race of bearing (15). Select spacer (17) of proper thickness to limit gap to 0.080-0.300 mm (0.003-0.012 inch). Spacer (17) is available in four thicknesses of 4.14-4.19, 4.39-4.44, 4.62-4.67 and 4.85-4.90 mm (0.163-0.165, 0.173-0.175, 0.182-0.184 and 0.191-0.193 inch). Reassemble mainshaft and components inside housing using the preselected spacer (17). Flat side of spacer (17) should be toward gear (13).

Remainder of assembly is reverse of disassembly procedure. Coat threads of screws attaching retainer housing (3) lightly with "Hylomar" or equivalent sealer before tightening to 54-61 N·m (40-45 ft.-lbs.) torque.

COUNTERSHAFT (LAYSHAFT) AND GEARS

Models With Eight-Speed Shuttle Transmission

157. The mainshaft must be removed as outlined in paragraph 156 and pto lower shaft (46—Fig. 187) and drive gear (45) should be removed as outlined in paragraph 154, before removing countershaft. A $7/16$ inch UNF screw approximately 75-100 mm (3-4 inches) long can be threaded into front of pto lower shaft and used to assist in removal of lower pto shaft. Remove snap ring (25) from rear, bump shaft forward enough to remove snap ring (29) from groove and move snap ring forward to unsplined part of countershaft. Remove snap ring (35) and thrust washer (84). Drive countershaft (31) to the rear and slide reverse gear (34) forward as far as possible. Remove snap ring (86) from groove, then slide snap ring forward as far as possible. Drive countershaft (31) toward rear as far as possible and remove reverse gear (34). Remove snap ring (86) and synchronizer ring (75), then remove synchronizer assembly (76 through 79) as a unit. Slide synchronizer ring (80), synchronizer cone (85), gear (33) and washer (88) from shaft. Drive shaft and bearing (32) forward through center web, then slide shaft (31), snap ring (87) and bearing (32) out toward front and remove gears (27, 28 and 69).

Synchronizer parts (76 through 79) should be removed as an assembly and separated only if necessary. Wrap synchronizer assembly with a cloth to catch poppet parts (77 and 78) when coupling (76) is removed from hub (79). Clean, inspect and renew any components that are damaged or worn excessively.

Place synchronizer cones (74 and 85) and synchronizer rings (75 and 80) on respective gears (33 and 34). Make sure that each ring is seated squarely, then measure clearance between ring and gear at several places around gear, using a feeler gauge. Install new synchronizer ring if clearance is less than 0.5 mm (0.020 inch). This clearance can be checked with gearbox assembled, but minimum clearance is increased to 0.8 mm (0.030 inch).

When reassembling the synchronizer shift components, align centralizing holes (H—Fig. 188) in shift collar (76) and hub (79). Synchronizer rings (75 and 80) on countershaft are different than similar parts on mainshaft, because countershaft rotates in opposite direction. Special tools (MF 415 and MS 550) are available to assist holding the three poppets (77 and 78) into hub (79) while installing coupling (76).

Reassemble by reversing removal procedure. Always install new snap rings when reassembling. Position snap ring (29) around shaft in the unsplined area and install front bearing (32) on shaft before installing shaft in housing. Refer to Fig. 186 for cross section of assembled transmission. Refer to paragraph 156 for assembly of mainshaft and selection of thrust washer (17—Fig. 187).

Coat gasket (43—Fig. 185) lightly with "Hylomar" or equivalent sealer before installing. Coat threads of screws attaching cover (36) and screws attaching retainer housing (3—Fig. 184) lightly with "Hylomar"

or equivalent sealer before tightening to 54-61 N•m (40-45 ft.-lbs.) torque.

REVERSE IDLER GEAR AND SHAFT

Models With Eight-Speed Shuttle Transmission

158. To remove reverse idler gear (54—Fig. 187), shaft (57) and related parts, first remove the transmission as outlined in paragraph 149. Remove the top cover and shift levers as outlined in paragraph 150, planetary unit as outlined in paragraph 152 and the shift rails and forks as outlined in paragraph 151. Refer to paragraph 154 and remove the pto lower shaft front bearing and retainer, then refer to paragraph 155 and remove the transmission input shaft. Remove the spacer (83) from front of mainshaft.

Carefully withdraw reverse idler shaft (57) from front and lift idler gear (54), spacer (52), washers (51), thrust washers (50) and bearing rollers (53) through top opening, being careful not to drop any parts into housing. Each bearing path (53) contains 23 rollers.

Assemble needle rollers (53) and washers (51) in reverse idler gear using petroleum jelly. Position thrust washers (50), spacer (52) and gear and bearing assembly in housing, then install idler shaft (57), using new "O" ring (56). Complete reassembly by reversing disassembly procedure.

TRANSMISSION (TWELVE-SPEED SHUTTLE)

365, 375, 390, 390T and 398 Models

Twelve forward speeds are possible by using a synchromesh gearbox with four forward speeds compounded by a three-speed range unit. An equal number of reverse speeds is made possible by fitting a forward/reverse unit in the front of gearbox. Refer to Fig. 189 for shift pattern of twelve-speed shuttle transmission available in 365, 375, 390, 390T and 398 model tractors. The forward/reverse shift lever (1) is located to the left of the steering wheel and can be moved when the clutch pedal is depressed. The engine can be started only when forward/reverse lever is in neutral (center) position. Selection of the twelve speed ranges is accomplished by moving a single lever (2). The shift lever is spring loaded to the center, 3rd and 4th speed, position. A three speed range transmission is shifted by moving lever (2) to neutral position, then to the far right, before moving the ratcheting shift either up or down. An indicator (3) shows which range is selected. After range is selected,

the shift lever can be moved to select one of the four gears of the main transmission in a conventional manner.

LUBRICATION

Models With Twelve-Speed Shuttle Transmission

159. The transmission, rear differential and hydraulic system share common fluid which is contained in the transmission and center housing. Recommended lubricant is Massey-Ferguson Super 500 multi-use 10W-30 oil, Massey-Ferguson Permatran oil or equivalent transmission/hydraulic oil. Capacity is 47.4 L (12.5 gal.). Tractors are equipped with dipstick with marks indicating maximum and minimum oil levels. Level should be maintained at maximum level if tractor is operated on hilly terrain or if

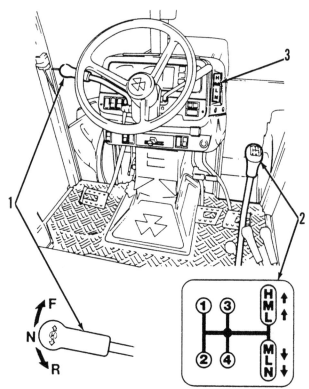

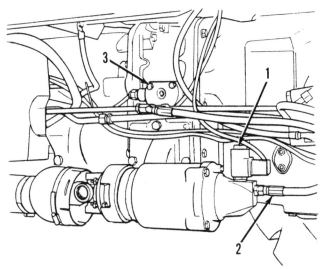

Fig. 190—View of left side showing hose (2) and solenoid (1) to operate the four-wheel drive. The range indicator switch (3) is retained by three screws.

Fig. 189—Shift pattern for twelve-speed shuttle transmission. The forward/reverse shift lever is located at (1) on left of steering wheel. Selection of the twelve speed ranges is accomplished by moving a single lever (2). Indicator (3) shows which of the three ranges is engaged.

using implements which require large quantities of oil.

TRACTOR REAR SPLIT

Models With Twelve-Speed Shuttle Transmission

160. Do not attempt to split tractor between the range unit and the main gearbox. The main gearbox and range gearbox should be separated only after they have been removed from the tractor as a unit.

To separate tractor between the rear of the transmission housing and the front of the rear axle center housing, set parking brake and block the wheels to prevent rolling. Disconnect battery ground cable and drain all transmission fluid.

If equipped with four-wheel drive, refer to paragraph 9 and remove the front drive shaft. Unscrew and remove the solenoid (1—Fig. 190) and detach hydraulic line (2). Detach four-wheel-drive indicator switch and four-wheel-drive selector lever.

On models with cab, disconnect throttle stop cable and throttle cable from fuel injection pump and move cables out of the way so that cables will not catch when separating tractor. Mark (identify), disconnect

and cap steering hoses, heater hoses and air conditioning lines (if so equipped), which would interfere. Remove front floor mat and floor plates.

On models without cabs, remove floor mats, floor panels, lower (gearbox) cover, both foot steps and battery boxes.

On all models, unbolt and withdraw range indicator unit (3), disconnect clutch linkage and remove foot throttle pedal. Remove knob (1—Fig. 191) from main shift lever, then remove the metal guard (2) from around shift linkage. Remove the lower strut (3) from the shift lever, remove the shift lever pivot roll pin (4), then withdraw shift lever (5). Disconnect hydraulic oil supply pipe located on right side of tractor and, if so equipped, detach hydraulic pipes from left side. Disconnect forward/reverse shift linkage and brake lines, then drain brake system. Disconnect wiring harness at the connectors below console on right side and disconnect wires from safety start switch. Remove the hydraulic suction filter assembly. Place hardwood wedges between the front axle and axle support casting on both sides to prevent tipping around axle pivot when tractor is separated. Support rear of tractor under axle center housing and front of tractor under transmission housing. Remove screws attaching rear axle center housing to four-wheel-drive transfer case, then separate the tractor halves. Refer to Fig. 192. Either the front can be moved forward or the rear can be moved to the rear, but it is important to support both halves safely and securely.

Install at least two dowel studs to facilitate alignment, position new gasket on dowel studs, then move tractor together, turning engine to align splines of shear tube. Make sure that flanges are tight against each other before installing retaining screws. Tighten all screws retaining the rear axle center housing to the transmission housing to 102-122 N·m (75-90 ft.-

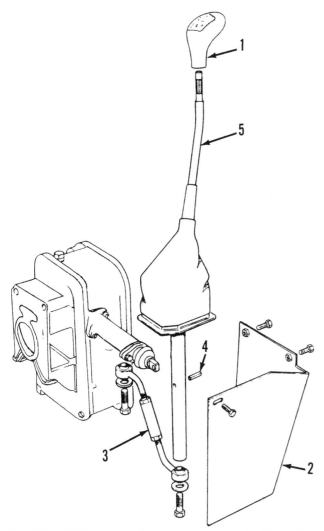

Fig. 191—Shift lever and related parts used with twelve-speed shuttle transmission.

1. Shift knob
2. Shield
3. Lower strut
4. Pivot roll pin
5. Shift lever

lbs.) torque, beginning at top center and progressing in a clockwise direction (as viewed from tractor rear) two times around the mating flange. If equipped with four-wheel drive, coat threads of drive shaft retaining screws with "Loctite 270," then attach drive shaft flanges and tighten screws to 55-75 N•m (40-55 ft.-lbs.) torque. Remainder of assembly is reverse of disassembly.

TRANSMISSION REMOVAL

Models With Twelve-Speed Shuttle Transmission

161. Refer to paragraph 160 and separate transmission from rear axle center housing, then support the gearbox assembly independently from the engine and front assembly. If not equipped with cab, unbolt and support or remove the steering and instrument console. On all models, remove screws and stud nuts attaching transmission housing to the rear of engine, then separate transmission from engine.

Reassemble transmission to engine by reversing the splitting procedure and observe the following. Install a M12 alignment stud approximately 100 mm (4 inches) long in each side of the transmission housing to assist in aligning engine to transmission. Turn flywheel to align clutch plate splines with transmission and pto input shaft splines. Install retaining screws and nuts after engine and transmission housing flanges are completely together. Tighten retaining screws and nuts to 100-130 N•m (74-94 ft.-lbs.) torque.

Refer to paragraph 160 and reconnect transmission to rear axle center housing. Tighten all screws retaining the transmission housing to spacer housing, transfer case or rear axle center housing to 102-122 N•m (75-90 ft.-lbs.) torque, beginning at top center and progressing in a clockwise direction (as viewed from tractor rear) two times around the mating flange. On four-wheel-drive models, coat threads of drive shaft retaining screws with "Loctite 270," then attach drive shaft flanges and tighten screws to 55-75 N•m (40-55 ft.-lbs.) torque. On all models, check and adjust clutch linkage as outlined in paragraph 92.

Refer to paragraph 20 for removal of the four-wheel-drive transfer case from the range housing.

TOP COVER

Models With Twelve-Speed Shuttle Transmission

162. If transmission is not already removed or the top is not yet uncovered, remove the cab floor or foot plates. Disconnect the forward/reverse shift linkage. If equipped with cab, remove the three screws from the oil filler tube and remove the tube (6—Fig. 193). On all models, unbolt and remove top cover from transmission housing.

Install new "O" ring (4) if forward/reverse shift lever (1) is removed. Use thread lock such as "Loctite 270" on threads of set screw (2).

Before installing top cover, coat sealing surface with "Loctite 515 Instant Gasket" or equivalent sealer. Tighten retaining screws to 101-122 N•m (75-90 ft.-lbs.) torque.

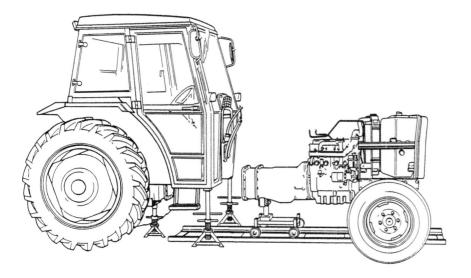

Fig. 192—View of tractor with standard cab separated between the transmission housing and the rear axle center housing. Special splitting stand and track is shown. Support and splitting procedures are similar for other models.

MAIN GEAR SHIFT LEVER AND SELECTOR SHAFT

Models With Twelve-Speed Shuttle Transmission

163. Remove knob (1—Fig. 191) from main shift lever, then remove the metal guard (2) from around

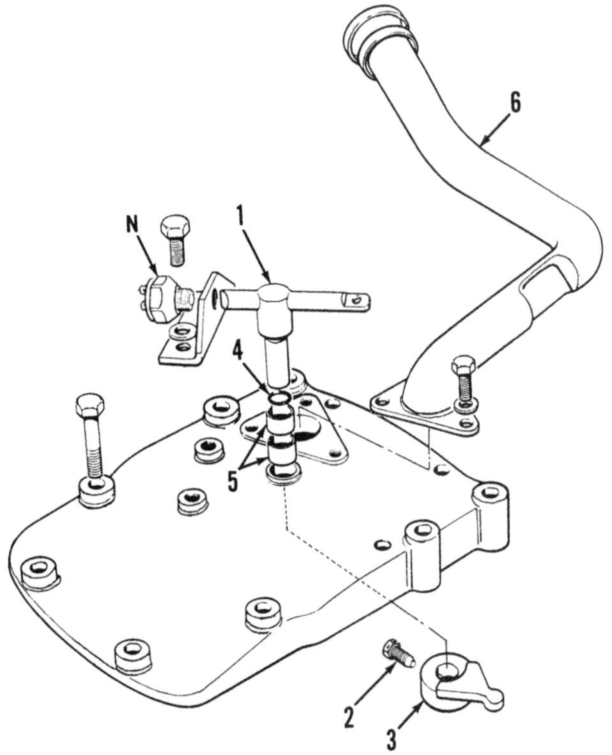

Fig. 193—View of transmission top cover and forward/reverse shift lever.

N. Neutral switch
1. Forward/reverse shift
2. Set screw
3. Internal shift lever
4. "O" ring
5. Bushings
6. Fill tube

shift linkage. Remove the lower strut (3) from the shift lever, remove the shift lever pivot roll pin (4), then withdraw shift lever (5).

Remove the two 12-point screws (11—Fig. 194) and withdraw the selector assembly (1 through 12) from the range housing.

Install new "O" ring (3) if unit is disassembled. Coat sealing surface of range housing and sealing surface of end cover (5) with "Loctite 515 Instant Gasket" or equivalent before installing. Tighten the 12-point screws (11) to 50-70 N·m (38-52 ft.-lbs.) torque.

FORWARD/REVERSE SHIFT CONTROL LINKAGE

Models With Twelve-Speed Shuttle Transmission

164. Refer to Fig. 195 and Fig. 196 for forward/reverse shuttle shift linkage. Be sure that electrical wiring does not touch any part of shift linkage. Lubricate linkage with general purpose grease, when assembling. Adjust lengths of rods attached to the relay lever as shown in Fig. 197. Install shims between relay lever pivot and retaining snap ring to remove any end play that would result in lost motion.

MAIN TRANSMISSION SHIFT RAILS AND FORKS

Models With Twelve-Speed Shuttle Transmission

165. To remove the shift rails and forks for the main transmission, first remove the gearbox from the tractor as outlined in paragraph 161 and remove the top cover as outlined in paragraph 162. Refer to paragraph 163 and remove the main gear shift lever and

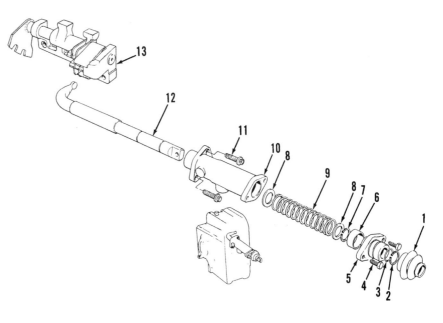

Fig. 194—Exploded view of the gear selector shaft and related parts.

1. Boot
2. Snap ring
3. "O" ring
4. Screws (6 point)
5. End cover
6. Spacer
7. Snap ring
8. Washers (2 used)
9. Spring
10. Support
11. Screws (12 point)
12. Selector shaft
13. Rear selector mechanism

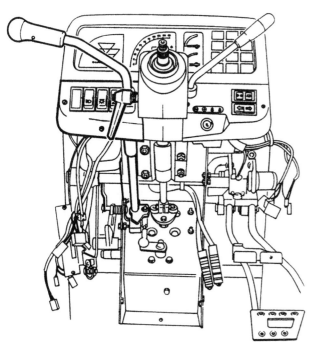

Fig. 195—View of forward/reverse shift lever and linkage installed, but with steering wheel and lower cover removed.

selector shaft. Set the transmission on front end, with engine/clutch flange down. Unbolt range transmission housing from the main transmission housing, then carefully lift range gearbox from main gearbox.

Unbolt and remove the rear selector mechanism (Fig. 198). Remove safety wire and set screws (5—Fig. 199) and lift detent springs and plungers (4) from bores in housing. Slide rail out of housing toward rear and remove gate and shift fork through top. Twist rails if necessary while withdrawing. Fork (8) can not be removed until transmission input shaft and gears are removed as outlined in paragraph 172.

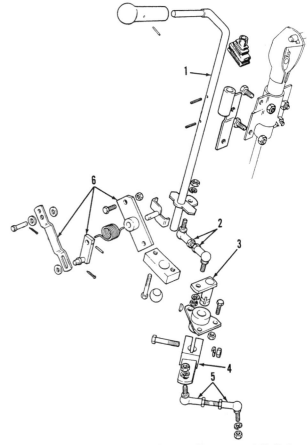

Fig. 196—Exploded view of forward/reverse shift linkage.

1. Shift lever	4. Lever
2. Link	5. Link
3. Relay shaft & lever	6. Clutch interlock linkage

When assembling, lubricate rails and make sure that rails do not have burrs, especially around the set screw locations, that would prevent easy and smooth installation. Each of the three shift rails is different

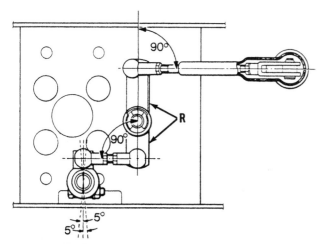

Fig. 197—Adjust lengths of shift control rods attached to the relay lever (R) to provide alignment of forward/reverse shift linkage as shown.

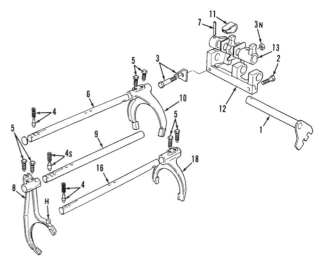

Fig. 199—Exploded view of shift forks. Detent spring and plunger (4S), in the center, is different than detents and springs (4) on either side.

1. Range selector shaft
2. Cap screw
3. Screw and plate
3N. Nut
4. Detent springs & plungers
5. Retaining screws
6. Shift rail (3rd/4th)
7. Roll pin
8. Shift fork
9. Shift rail (Forward/Reverse)
10. Shift fork
11. Return spring
12. Support
13. Gasket
14. Interlock plunger & balls
15. Retainer
16. Shift rail (1st/2nd)
17. Shift gate
18. Shift fork

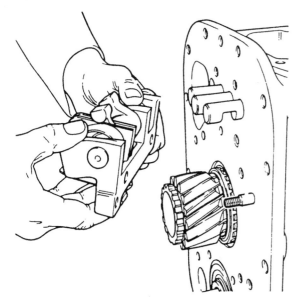

Fig. 198—The rear selector mechanism can be removed after removing the cap screw from the left and the nut from the right.

from the other rails. Adjust position of shift forks (8, 10 and 18) as follows.

Use special tool MF 414/2 or equivalent (T2—Fig. 200) to hold shift rail (9—Fig. 199) in neutral detent position. Insert centralizing tool, MF 414/1 or equivalent, (T1—Fig. 200) into hole (H—Fig. 199) in forward and reverse shift fork (8), into matching hole of sliding synchronizer coupling and into hole in synchronizer hub. Tighten each of the two set screws (5), while turning centralizing tool. Screws (5) should be tightened to 34-52 N·m (25-38 ft.-lbs.) torque and locked in position with safety wire.

To adjust the position of the 1st/2nd shift fork (18—Fig. 199) and 3rd/4th shift fork (10), proceed as follows. Use special tool MF 414/2 or equivalent (T2—

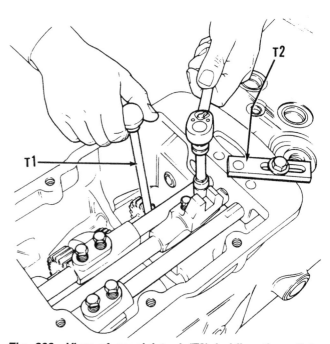

Fig. 200—View of special tool (T2) holding the rail in center detent position, while centering forward/reverse shift fork with special tool (T1).

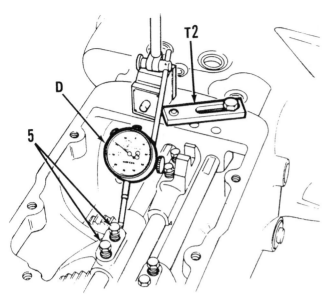

Fig. 201—View of special tool (T2) holding the rail in center detent position, while centering shift fork with dial indicator (D).

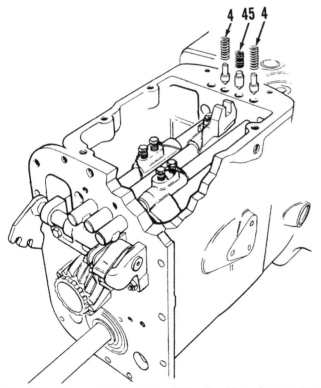

Fig. 202—View of detents for shift rails. Note that center detent and spring (4S) are different than the other two (4).

Fig. 201) to hold shift rails (6 and 16—Fig. 199) in neutral detent position. Make sure that synchronizer is in neutral and is not engaging either gear, then mount a dial indicator with a magnetic base as shown at (D—Fig. 201) against the shift fork. Gently move the shift fork to the rear, then forward without engaging either gear, while noting the location of the fork

as indicated by the dial indicator (D). Position the fork in exact center and tighten each of the two set screws (5). Screws should be tightened to 34-52 N•m (25-38 ft.-lbs.) torque and locked in position with safety wire. Make sure that tightening the screws does not move the position of the fork on rail. Use similar procedure to adjust other fork (10 or 18—Fig. 199) on rail (6 or 16).

Shift into each gear to make sure that each can be correctly engaged. Shift all gears, rails and gates to neutral after checking to make sure that all shift gates align. Install detent plungers and springs (4 and 4S—Fig. 202). Remainder of assembly is the reverse of disassembly procedure.

Reattach range gearbox to main transmission, using a new gasket and coating threads of retaining screws with "Hylomar" or equivalent sealer. After lowering range gearbox onto main gearbox, make sure that range shift fork is correctly engaged. Tighten screws attaching range gearbox to main gearbox to 105 N•m (75 ft.-lbs.) torque. Refer to paragraph 167 for selecting shims (40—Fig. 203) as required for setting shaft end play.

RANGE CHANGE UNIT UPPER SHAFT

Models With Twelve-Speed Shuttle Transmission

166. To remove the upper shaft from the range transmission, first remove the gearbox from the tractor as outlined in paragraph 161. Remove the main gear shift lever and selector shaft as outlined in paragraph 163, unbolt range transmission housing from the main transmission housing, then separate the two housings. Support the transmission on a suitable stand with rear of range gearbox down and enough clearance to withdraw shaft (37—Fig. 203). Remove screws (42) from front of selector rail, remove screw (43) from rear, then withdraw rail (45) from shifter forks (46 and 47). Support gears and forks using special tool MF477 (P2—Fig. 204) or equivalent. Attach a fixture such as MF476 (P1) and push the upper drive shaft out of bearing (14—Fig. 203).

> **CAUTION: Do not drop parts as shaft is withdrawn from bearing. Also, do not use a hammer and punch to remove the shaft from bearing or to install bearing on shaft, because serious, but undetected damage will occur.**

Remainder of disassembly procedure will be evident. Pins, springs and retainers (25—Fig. 203) can be driven from gears (24 and 35) as shown in Fig. 205 if renewal is required. Refer to paragraph 169 for removal and service to shifter cam assembly (49—Fig. 203).

Assemble synchronizer cone (28 or 34—Fig. 203) and synchronizer ring (29 or 32) on appropriate gear (24 or 35) and measure gap with feeler gauge as shown in Fig. 206. If average clearance is less than 0.5 mm (0.020 inch), install new brass synchronizer ring.

Use petroleum jelly to hold synchronizer cones, rings, bearings and other parts together while assembling. Install low range gear (35—Fig. 203), bushing (36) and related parts in place, using special holding fixture MF477 or equivalent (P2—Fig. 204). "V" notches in synchronizer cone (34—Fig. 203) must

engage drive pins (25) in gear (35). Assemble Middle/Low range synchronizer (29 through 32) and shift fork (47) over Low drive gear (35). Pins and rollers (48) must be properly located on shift fork and must correctly engage track in cam (49). Be sure that locating lugs of synchronizer ring (29) engage slots of hub (30). Install Middle range gear (24) and bushing (23), making sure that "V" notches in synchronizer cone (28) engage drive pins (25) of gear (24). Install four-wheel-drive gear (22) with grooved thrust face down toward gear (24). Assemble synchronizer (16 through 21) and High range shift fork (46) into posi-

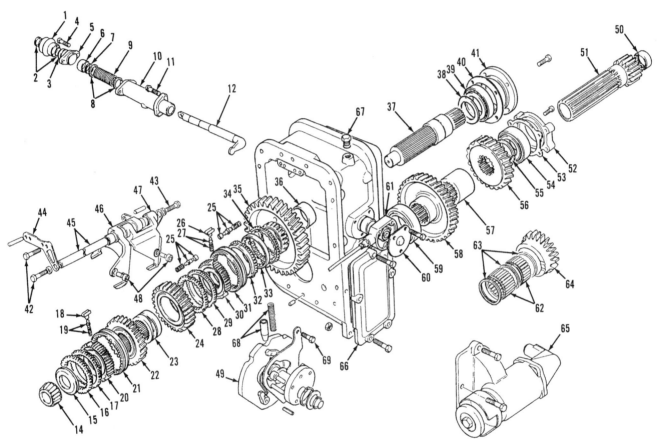

Fig. 203—Exploded view of range section of twelve-speed shuttle transmission. Three sets of pins, springs and retainers (25) are located in each gear (24 and 34).

1. Boot
2. Snap rings
3. "O" ring
4. Screws (6 point)
5. End cover
6. Spacer
7. Snap ring
8. Washers (2 used)
9. Spring
10. Support
11. Screws (12 point)
12. Selector shaft
14. Tapered roller bearing
15. Thrust washer
16. High range synchronizer cone
17. Synchronizer ring
18. Pressure blocks

19. Springs and plungers
20. Synchronizer hub
21. Shift collar (High range)
22. Drive gear (four-wheel drive)
23. Bushing
24. Middle range gear
25. Alignment pins, springs & retainers
26. Pressure blocks
27. Springs and plungers
28. Synchronizer cone
29. Synchronizer ring
30. Synchronizer hub
31. Shift collar (Medium/Low range)
32. Synchronizer ring
33. Synchronizer bushing

34. Synchronizer cone
35. Low range gear
36. Bushing
37. Upper drive shaft
38. Thrust washer
39. Tapered roller bearing
40. Shims
41. Bearing cap
42. Screws
43. Screw
44. Plates (2 used)
45. Selector rail
46. Shift fork (High range)
47. Shift fork (Middle/Low range)
48. Rollers and pins
49. Shifter cam assembly
50. Bearing

51. Lower shaft
52. Bearing cap
53. Shims
54. Tapered roller bearing
55. Thrust washer (2 pieces)
56. Middle range gear
57. Spacer
58. Driven gear
59. Tapered roller bearing
60. Indicator switch cover
61. Range indicator switch
62. Roller bearings
63. Spacer rings
64. Drive gear (four-wheel drive)
65. Transfer case (four-wheel drive)
66. Cover (two-wheel drive)

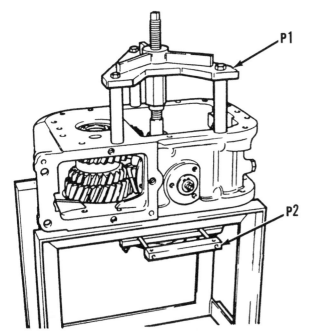

Fig. 204—View of special tool MF476 (P1) attached for pushing the upper drive shaft out of front tapered bearing. Special tool MF477 (P2) is available to hold gears and synchronizers in position while removing and installing shaft.

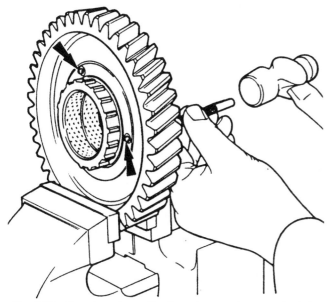

Fig. 205—Synchronizer aligning pins, springs and retainers can be removed by driving from gears with a punch as shown.

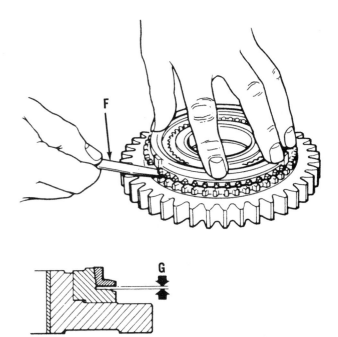

Fig. 206—Measure gap (G) with a feeler gauge (F) as shown to check condition of synchronizer rings.

of shims (40) will require later removal of cap. Install thrust washer (15) with smaller diameter up (toward front), then press tapered roller bearing (14) onto end of shaft.

> **CAUTION: Do not use a hammer and punch to drive bearing onto shaft. Damage resulting in seizure and extensive failure to many components may occur.**

A special pressing fixture MF476 (P1—Fig. 204) is available to assist in removal and installation of bearing (14—Fig. 203). Make sure that bearing is fully in place by attempting to turn the thrust washer (15). Install shifter rail (45) and screws (42 and 43). Coat threads of screws (42 and 43) with "Loctite 270" or equivalent and tighten to 25-35 N·m (18-26 ft.-lbs.) torque. Rotate shafts (37 and 51) to make sure that all gears are free to rotate and are not binding.

Reattach range gearbox to main transmission, using a new gasket and coating threads of retaining screws with "Hylomar" or equivalent sealer. After lowering range gearbox onto main gearbox, make sure that range shift fork is correctly engaged. Tighten screws attaching range gearbox to main gearbox to 105 N·m (75 ft.-lbs.) torque. Refer to paragraph 167 for selecting shims (40) as required for setting shaft end play.

tion. Make sure that pins and rollers (48) correctly engage track in cam (49) and bushing bores in shift forks are aligned with bore for shift rail (45). Assemble thrust washer (38), bearing (39) and cap (41) without shims (40) and insert shaft up through the installed gears, bushings, hubs and related parts. Install screws to hold cap (41) in place, but selection

RANGE CHANGE UPPER SHAFT BEARING PRELOAD

Models With Twelve-Speed Shuttle Transmission

167. The selection of shims (40—Fig. 203) determines the adjustment of bearings (14 and 39) and proper adjustment of these bearings is critical to longevity of transmission parts. Some special tools are required. The special rotating tool (T3—Fig. 207), which can be locally fabricated from an old shear tube or pipe and handle, is used to turn the transmission shaft before measuring gap with feeler gauges (F). Special tool MF478 (T4) is four special length M6 screws and four special springs which are used to load cap (41) to about 45 kg. (100 lbs.).

To check bearing preload, set transmission on floor with engine/clutch surface down and remove the four screws retaining bearing cap (41—Fig. 207) to the range gearbox housing. Install four M6 screws with springs (T4) and tighten the screws until the combined pressure of the four springs is approximately 45 kg. (100 lbs.) and gap between cap and housing is even. If special screws and springs MF478 are used, height (H) between top of screw heads and bearing cap is 55 mm (2.17 inches). Engage medium range by moving the range rear shift coupling down (toward front), then turn the upper shaft with tool (T3) approximately 10 revolutions to make sure that bearings are correctly seated. Measure clearance between bearing cap and gearbox housing carefully in at least four locations around cap. If measurements are not the same around cap, rotate upper shaft again with

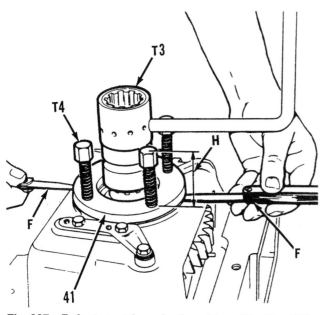

Fig. 207—Refer to text for selecting shims (40—Fig. 203). Special tool (T4) is four M6 screws and four springs.

special tool and make sure that special tool springs are all compressed the same amount, then remeasure. Subtract approximately 0.05 mm (0.020 inch) from the gap measured with feeler gauges, then install shims (40—Fig. 203) equal to this thickness. An example is a measurement of 0.37 mm - (minus) 0.05 mm equals 0.32 mm. Shims are available in thicknesses of about 0.5, 0.20 and 0.50 mm (0.002, 0.008 and 0.020 inch). Be sure to measure thickness of shims selected. Slightly less preload is required if original bearings are reinstalled.

Remove the special screws, springs, turning handle and bearing cap, then install shims determined to be correct thickness. Install bearing cap, coat threads of retaining screws with "Loctite 270" or equivalent and tighten retaining screws to 25-35 N·m (18-26 ft.-lbs.) torque. Disengage middle range gear and check both shafts for free rotation.

RANGE CHANGE LOWER SHAFT

Models With Twelve-Speed Shuttle Transmission

168. Refer to paragraph 166 and remove the upper shaft. Unbolt and remove the lower bearing cap (52—Fig. 203), shims (53) and cup for bearing (54). Withdraw lower shaft (51), cone for bearing (54) and thrust washer (55), then lift middle range gear (56), spacer (57), lower shaft drive gear (58) and if so equipped, the four-wheel-drive gear (64). Press cone of bearing (54) toward rear (gear end of shaft) to remove the two piece thrust washer (55).

Install cup for bearing (59) in housing bore, then install drive gear (58) with cone for bearing (59). On four-wheel-drive models, assemble spacer rings (63), bearings (62) and gear (64) over spacer (57), then position the unit in housing against gear (58). On two-wheel-drive models, install spacer (57) against gear (58). On all models, press bearing (50) into lower shaft and press cone for bearing (54) onto shaft. Position both halves of thrust washer (55) in groove of shaft, then press cone of bearing (54) toward front of shaft over thrust washer. Install middle range gear (56), then insert the lower shaft (51) through gears and spacers. Install bearing cap (52) and cup for bearing (54), using 1.0 mm (0.039 inch) thickness of shims (53). Shims are available in 0.05, 0.20 and 0.50 mm (0.002, 0.008 and 0.020 inch) thicknesses. Tighten screws retaining cap (52) to 25-35 N·m (18-26 ft.-lbs.) torque and mount a dial indicator as shown in Fig. 208 to measure shaft end play. Bearing preload should be 0.10 mm (0.004 inch) for new bearings or 0.05 mm (0.002 inch) for used bearings. Unbolt and remove cap (52), then reduce thickness of shims the amount of the measured end play plus the additional amount of desired preload. Reinstall cap with the

Fig. 208—Mount a dial indicator (D) as shown, then pry gear up (P) to check end play of lower shaft in bearings.

required amount of shims, coat threads of retaining screws with "Loctite 270" or equivalent and tighten screws to 25-35 N·m (18-26 ft.-lbs.) torque. Check shaft for free rotation, refer to paragraph 166 to

install the upper shaft and to paragraph 167 for setting upper shaft bearings.

RANGE SELECTOR CAM

Models With Twelve-Speed Shuttle Transmission

169. Refer to paragraph 166 and remove the upper shaft and selector forks. Remove cover (60—Fig. 203) and indicator switch (61), if not already removed. Remove plug (67), spring and plunger (68). Remove screw (69) and withdraw cam assembly (49) from housing. Refer to Fig. 209 for disassembled view of switch and cam assembly. Further disassembly is possible after removing snap ring.

Refer to Fig. 210 for correct assembly of pawls (79) and spring plate (77). Stick rollers (76) to cam block (75) with petroleum jelly, while assembling. Use a small screwdriver or similar tool to carefully guide pawls (79) into position so that pins (P) will enter holes in pawls. Coat threads of screw (69—Fig. 209) with "Loctite 270" or equivalent and tighten to 25-35 N·m (18-26 ft.-lbs.) torque. Install plunger and spring (68) and plug (67) making sure that selector cam is in neutral position. Tighten plug (67) to 50-70 N·m

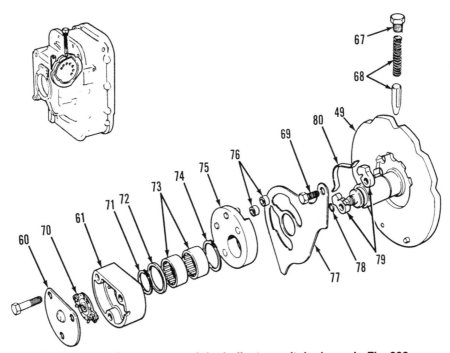

Fig. 209—Exploded view of the range selector cam and the indicator switch shown in Fig. 203.

49. Shifter cam			
60. Indicator switch cover	69. Screw	73. Needle bearings	77. Spring plate
61. Indicator switch housing	70. Indicator switch	74. Snap ring	78. "O" ring
67. Plug	71. Snap ring	75. Cam block	79. Pawls
68. Spring and detent	72. Thrust washer	76. Rollers	80. Spring

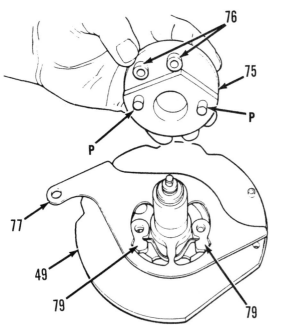

Fig. 210—*View of selector cam, spring plate, ratchet pawls and spring correctly assembled.*

(38-52 ft.-lbs.) torque. Refer to paragraph 166 for installation of the shift forks and upper shaft.

PTO INPUT SHAFT AND RETAINER HOUSING

Models With Twelve-Speed Shuttle Transmission

170. To remove the pto input shaft (8—Fig. 211) and retainer (3), first split tractor between engine and transmission as outlined in paragraph 93. Remove clutch release bearing (16), retainer (15), fork (L) and release shaft (S). Remove brake cross shaft, if so equipped. Remove screws and pull pto input shaft (8) and retainer (3) from transmission housing as a unit.

To disassemble, remove snap ring (6) from rear of retainer, then push input shaft and bearing (5) out of housing toward rear. Bearing can be removed from input shaft after removing snap ring (4). Remove oil

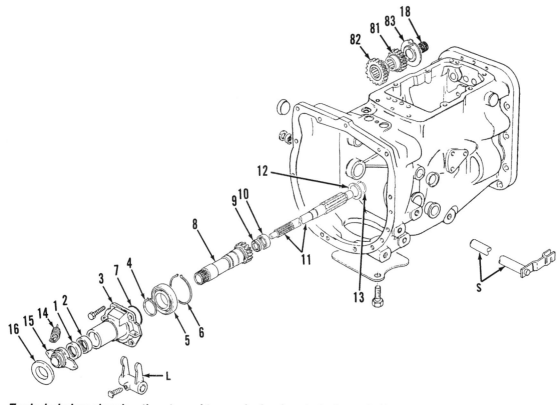

Fig. 211—*Exploded view showing the pto and transmission input shafts and related parts. Clutch release lever is shown at (L) and lever shaft is shown at (S). A spacer is used in place of gear (82) on 398 models and smaller tractors with high speed shuttle transmission.*

L. Clutch release lever		15. Release housing
S. Clutch lever and shaft	5. Ball bearing	16. Release bearing
1. Oil seal	6. Snap ring	10. Needle bearing	18. Needle bearing
2. Needle bearing	7. "O" ring	11. Transmission input shaft	81. Forward gear
3. Retainer housing	8. Pto input shaft	12. Thrust washer	82. Reverse gear
4. Snap ring	9. Oil seal	13. Snap ring	83. Spacer
		14. Return spring	

seals (1 and 9) and needle bearing (2) only if new parts are to be installed.

Lubricate all seals and bearings with petroleum jelly before assembling. Use special tool (MF 315A or equivalent) to install new needle bearing (2). Install seal (9) with lip toward rear, using special tool (MF 421 or equivalent). Install ball bearing (5) on shaft with shield toward rear, then install snap ring (4). Use special tool (MF255B or equivalent) to install seal (1) with lip toward rear. Use special sleeve (KMF 1004 or equivalent) to protect seal (1), slide input shaft into the retainer housing, then install snap ring (6). Install "O" ring (7) and install housing and shaft assembly over transmission input shaft (11) into housing bore. Coat threads of retaining screws with "Hylomar" or equivalent sealer and tighten to 54-61 N·m (40-45 ft.-lbs.) torque. Remainder of assembly is reverse of disassembly.

PTO LOWER SHAFT FRONT BEARING AND RETAINER

Models With Twelve-Speed Shuttle Transmission

171. The pto lower shaft front bearing (41—Fig. 212) can be removed and installed after separating tractor between the rear of the engine and front of transmission housing. Removal of the lower driven gear (45) and shaft (46) require the additional steps of separating at rear of transmission housing and removal of the transmission top cover and shift rails. The following procedure describes only service to the front bearing and bearing retainer (42) after splitting as outlined in paragraph 93.

Remove clutch release bearing (16—Fig. 211), retainer (15), fork (13) and release shaft (12). Unbolt and remove cover (36—Fig. 212), then remove snap ring (38) and washer (39). Thread a ⅜ inch UNC cap screw, 75 mm (3 inches) long into each of the two threaded holes of retainer (42). Tighten the two jackscrews to push retainer (42) and bearing (41) forward off shaft (46). Bearing can be pressed from retainer after removing snap ring (40). Remove "O" ring (37) and gasket (43), then clean all parts thoroughly.

Coat gasket (43) and threads of retaining screws lightly with "Hylomar" or equivalent sealer before installing. End of shaft (46) is threaded to accept a puller screw to pull bearing (41) onto shaft until washer (39) and snap ring (38) can be installed. Align holes in retainer (42) with holes in transmission housing, install "O" ring (37) and cover (36). Tighten retaining screws to 54-61 N·m (40-45 ft.-lbs.) torque. Remainder of assembly is reverse of disassembly procedure.

MAIN TRANSMISSION INPUT SHAFT

Models With Twelve-Speed Shuttle Transmission

172. To remove the transmission input shaft (11—Fig. 211), first remove the transmission as outlined in paragraph 161. Remove the top cover and shift levers as outlined in paragraph 162 and the shift rails and forks as outlined in paragraph 165. Refer to paragraph 170 and remove the pto input shaft and retainer. Slide input shaft (11—Fig. 211) forward, then withdraw gears (81 and 82) and spacer (83) out top. Be careful that bearing (18) does not fall into transmission when withdrawing shaft (11). A spacer is installed in place of gear (82) on 398 models and smaller tractors with high speed transmission. Forward/reverse shift fork (8—Fig. 199) can be removed after removing input gears.

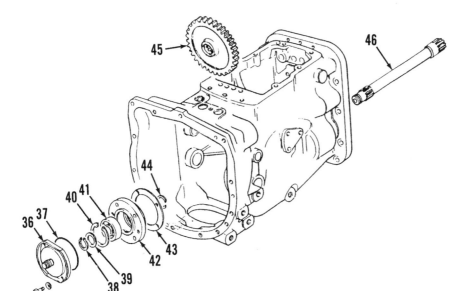

Fig. 212—Exploded view of the pto lower shaft and bearings.

36. Cover
37. "O" ring
38. Snap ring
39. Washer
40. Snap ring
41. Ball bearing
42. Retainer
43. Gasket
44. Snap ring
45. Pto lower gear
46. Pto lower shaft

Reinstall input shaft by reversing removal procedures. Refer to paragraph 165 for correctly setting position of forks on shift shafts.

MAIN TRANSMISSION UPPER SHAFT AND GEARS

Models With Twelve-Speed Shuttle Transmission

173. To remove the mainshaft (19—Fig. 213), first remove the transmission as outlined in paragraph 161. Remove the top cover and shift levers as outlined in paragraph 162 and the shift rails and forks as outlined in paragraph 165. Refer to paragraph 170 and remove the pto input shaft and retainer, then remove the transmission input shaft as outlined in paragraph 172.

Lift snap ring (47—Fig. 213) from groove, push mainshaft (19) to rear and remove gear (13). Continue to move mainshaft toward rear while lifting synchronizer assembly (90 through 95), thrust washer (61), gears (20, 21 and 24) and synchronizer assembly (63 through 68) out top opening. Rear bearing (22) can be removed if necessary.

Synchronizer parts (64 through 67 and 90 through 95) should be removed as an assembly and separated only if necessary. Wrap synchronizer assembly with a cloth to catch poppet parts (65 and 66 or 77 and 78) when sliding coupling (64 or 76) is removed from hub (67 or 79). Clean, inspect and renew any components that are damaged or worn excessively. Place synchronizer cones (72 and 73 or 89 and 96) and synchronizer rings (63 and 68 or 90 and 95) on respective gears (21 and 24 or 13 and 20). Make sure that each ring is seated squarely on taper of cone. Measure clearance between ring and gear at several places around gear, using a feeler gauge. Install new synchronizer ring (63, 68, 90 or 95) if clearance is less than 0.5 mm (0.020 inch). This clearance can be checked with gearbox assembled, but minimum clearance is then increased to 0.8 mm (0.030 inch).

When reassembling the synchronizer shift components, align centralizing holes (H—Fig. 214) in shift collar (64 or 91) and hub (67 or 92). Special tools (MF 415 and MS 550) are available to assist holding the three poppets (65 or 93) into hub (67 or 92), against spring pressure, while installing coupling (64 or 91). Synchronizer rings (75 and 80) on countershaft are different than similar parts on mainshaft, because countershaft rotates in opposite direction.

Make sure that forward and reverse shift fork (8—Fig. 199) is in place in shift collar before installing the mainshaft.

Reassemble mainshaft (19—Fig. 213), snap ring (47), bearings (48 and 22) and all parts between bearings (22 and 48), outside of gearbox housing. Be sure that sleeves (59, 60, 62 and 70) match respective gears (13, 20, 21 and 24) in length. Refer to Fig. 215 for cross section of assembled transmission. Measure clearance between front snap ring (47—Fig. 213) and inner race of bearing (48). Select spacer (49) of proper thickness to limit gap to 0.080-0.300 mm (0.003-0.012 inch). Spacer (49) is available in four thicknesses of 3.38-3.43, 3.63-3.68, 3.89-3.94 and 4.14-4.19 mm (0.133-0.135, 0.143-0.145, 0.153-0.155 and 0.163-0.165 inch). Reassemble mainshaft and components inside housing using the preselected spacer (49). Flat side of spacer (49) should be toward gear (13).

Remainder of assembly is reverse of disassembly procedure. Coat threads of screws attaching retainer housing (3—Fig. 211) lightly with "Hylomar" or equivalent sealer before tightening to 54-61 N·m (40-45 ft.-lbs.) torque.

MAIN TRANSMISSION COUNTERSHAFT (LAYSHAFT) AND GEARS

Models With Twelve-Speed Shuttle Transmission

174. The upper (mainshaft) must be removed as outlined in paragraph 173 and pto lower shaft (46—Fig. 212) and drive gear (45) should be removed as outlined in paragraph 171, before removing countershaft. A 7/16 inch UNF screw approximately 75-100 mm (3-4 inches) long can be threaded into front of pto lower shaft and used to assist in removal of lower pto shaft.

Remove snap ring (25—Fig. 213) and spacer (30) from rear of countershaft, bump shaft forward enough to remove split ring (29) from groove. Remove snap ring (35) and thrust washer (84), then push countershaft (31) to the rear forward as far as possible. Slide constant mesh gear (34) forward as far as possible, lift snap ring (86) from groove and slide snap ring forward. Push countershaft (31) toward rear and remove reverse gear (34) with synchronizer cone (74). Remove snap ring (86) and synchronizer ring (75), then remove synchronizer assembly (76 through 79) as a unit. Slide synchronizer ring (80), synchronizer cone (85), gear (33) and washer (88) from shaft. Push countershaft (31), bearing (32), snap ring (87) and bearing (32) out toward front, through center web, then remove gears (27 and 28).

Synchronizer parts (76 through 79) should be removed as an assembly and separated only if necessary. Wrap synchronizer assembly with a cloth to catch poppet parts (77 and 78) when coupling (76) is removed from hub (79). Clean, inspect and renew any components that are damaged or worn excessively. Place synchronizer cones (74 and 85) and synchronizer rings (75 and 80) on respective gears (33 and

34). Make sure that each ring is seated squarely, then measure clearance between ring and gear at several places around gear, using a feeler gauge. Install new synchronizer ring if clearance is less than 0.5 mm

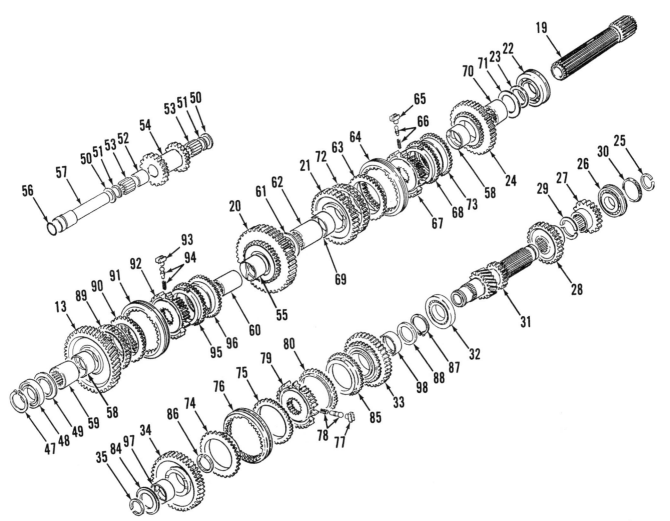

Fig. 213—Exploded view of twelve-speed shuttle transmission. Synchronizer (63 through 68) is made up of same components as synchronizer (75 through 80).

13. Gear (2nd)	35. Snap ring	64. Shifting coupler	79. Synchronizer hub
19. Mainshaft	47. Snap ring	65. Pressure blocks (3 used)	80. Synchronizer ring
20. Gear (1st)	48. Bearing	66. Ball pins & springs	(Same as 75)
21. Gear (4th)	49. Spacer (Variable	(3 used)	84. Thrust washer
22. Tapered roller	thicknesses)	67. Synchronizer hub	85. Synchronizer cone
bearing	50. Thrust washers	68. Synchronizer ring	86. Snap ring
23. Spacer	51. Washers	(Same as 63)	87. Snap ring
24. Gear (3rd)	52. Spacer	69. Bushing	88. Thrust washer
25. Snap ring	53. Needle rollers (23/path)	70. Sleeve	89. Synchronizer cone
26. Ball bearing	54. Gear (Reverse idler)	71. Washer	90. Synchronizer ring
& snap ring	55. Bushing	72. Synchronizer cone	91. Shifting coupler
27. Gear (3rd)	56. "O" ring	73. Synchronizer cone	92. Hub
28. Gear (4th)	57. Idler shaft	74. Synchronizer cone	93. Pressure blocks (3 used)
29. Spacer (2 piece)	58. Bushings	75. Synchronizer ring	94. Ball pins & springs
30. Spacer	59. Sleeve	(Same as 80)	(3 used)
31. Countershaft with	60. Sleeve	76. Shifting coupler	95. Synchronizer ring
1st & 2nd gears	61. Washer	77. Pressure blocks (3 used)	96. Synchronizer cone
32. Roller bearing	62. Sleeve	78. Ball pins & springs	97. Bushing
33. Forward gear	63. Synchronizer ring	(3 used)	98. Bushing
34. Reverse gear	(Same as 68)		

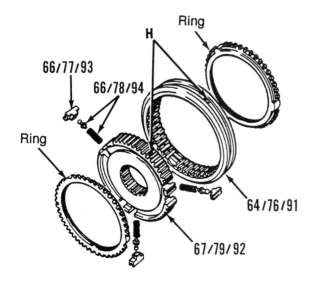

Fig. 214—Synchronizer rings for unit used on countershaft are different than similar parts used on mainshaft. Holes (H) are used when checking and adjusting position of shift forks and must be aligned when assembling synchronizer. Refer to Fig. 213 for legend.

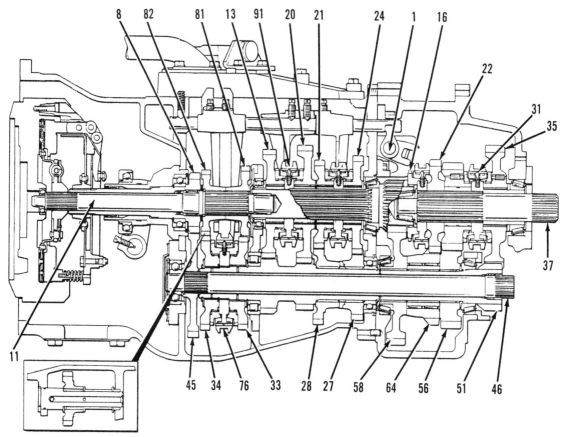

Fig. 215—Cross section of twelve-speed shuttle transmission.

1. Range selector shaft (Fig. 199)
8. Pto input shaft (Fig. 211)
11. Transmission input shaft (Fig. 211)
13. Second gear (Fig. 213)
16. High range synchronizer cone (Fig. 203)
20. First gear (Fig. 213)

21. Fourth gear (Fig. 213)
22. Front-wheel-drive gear (Fig. 203)
24. Third gear (Fig. 213)
27. Third gear (Fig. 213)
28. Fourth gear (Fig. 213)
31. Medium/Low range shift collar (Fig. 203)

33. Forward gear (Fig. 213)
34. Reverse gear (Fig. 213)
35. Low range gear (Fig. 203)
37. Upper range shaft (Fig. 203)
45. Pto lower gear (Fig. 212)
46. Pto lower shaft (Fig. 212)
51. Lower range shaft (Fig. 203)

56. Middle range gear (Fig. 203)
58. Driven gear (Fig. 203)
64. Front-wheel-drive gear (Fig. 203)
76. Shifting coupler (Fig. 213)
81. Forward gear (Fig. 211)
82. Reverse gear (Fig. 211)
91. Shifting coupling (Fig. 213)

(0.020 inch). This clearance can be checked with gearbox assembled, but minimum clearance is increased to 0.8 mm (0.030 inch).

When reassembling the synchronizer shift components, align centralizing holes in shift collar (76) and hub (79). Synchronizer rings (75 and 80) on countershaft are different than similar parts on mainshaft, because countershaft rotates in opposite direction. Special tools (MF 415 and MS 550) are available to assist holding the three poppets (77 and 78) into hub (79) while installing coupling (76).

Reassemble by reversing removal procedure. Always install new snap rings when reassembling. Position snap ring (29) around shaft in the unsplined area and install front bearing (32) on shaft before installing shaft in housing. Refer to Fig. 215 for cross section of assembled transmission. Refer to paragraph 173 for assembly of mainshaft and selection of thrust washer (49—Fig. 213).

Coat gasket (43—Fig. 212) lightly with "Hylomar" or equivalent sealer before installing. Coat threads of screws attaching cover (36) and screws attaching retainer housing (3—Fig. 211) lightly with "Hylomar" or equivalent sealer before tightening to 54-61 N·m (40-45 ft.-lbs.) torque.

REVERSE IDLER GEAR AND SHAFT

Models With Twelve-Speed Shuttle Transmission

175. To remove reverse idler gear (54—Fig. 213), shaft (57) and related parts, first remove the transmission as outlined in paragraph 161. Remove the top cover and shift levers as outlined in paragraph 162 and the shift rails and forks as outlined in paragraph 165. Refer to paragraph 170 and remove the pto lower shaft front bearing and retainer, then refer to paragraph 172 and remove the transmission input shaft.

Carefully withdraw reverse idler shaft (57) from front and lift idler gear (54), spacer (52), washers (51), thrust washers (50) and bearing rollers (53) through top opening, being careful not to drop any parts into housing. Each bearing path (53) contains 23 rollers. Some models have only a single gear at (54) and a spacer is installed.

Assemble needle rollers (53) and washers (51) in reverse idler gear using petroleum jelly. Position thrust washers (50), spacer (52) and gear and bearing assembly in housing, then install idler shaft (57), using new "O" ring (56). Complete reassembly by reversing disassembly procedure.

MAIN DRIVE BEVEL DRIVE GEARS AND DIFFERENTIAL

DIFFERENTIAL AND BEVEL RING GEAR

All Models

176. R&R AND OVERHAUL. The ring gear and differential unit can be removed from the differential housing after removing the rear axle housing assembly from the left side as outlined in paragraph 183.

To disassemble the removed differential unit, first scribe alignment marks on both halves of differential case to facilitate correct reassembly. Pull cone of bearing (4—Fig. 216) from differential lock coupler cap (8), then remove cap screws (7). Remove coupler cap and separate differential case halves (9 and 16). Differential pinions (13), side gears (11) and cross (12) can then be removed.

The main drive bevel ring gear (15) is secured to differential case half with either special bolts and nuts (18) or with rivets. Note the ring gear is a press fit on case and may also be epoxy bonded to case if previously serviced.

Inspect all parts for scoring, chipping, wear or other damage and renew if necessary. Backlash between pinion gears (13) and side gears (11) should be 0.08-0.28 mm (0.003-0.011 inch). It is suggested that cross (12) and all pinion gears (13) be renewed together.

To reassemble differential, reverse the disassembly procedure while noting the following special instructions. If bevel ring gear (15) was removed, install using special epoxy bonding kit. Be sure to follow kit instructions and cure epoxy for the correct time at the specified temperature. Apply "Loctite 270" or equivalent to threads of ring gear retaining screws, then tighten retaining nuts (18) evenly and to 160 N·m (120 ft.-lbs.) torque. Be sure to align previously affixed scribe marks on differential case and tighten retaining cap screws (7) to 108 N·m (80 ft.-lbs.) torque.

To reinstall differential, reverse the removal procedure. Refer to paragraph 177 if differential case halves and/or carrier bearings were renewed.

177. CARRIER BEARING PRELOAD. Differential carrier bearing preload is adjusted by installing the correct thickness spacer shield (3—Fig. 216) between bore in carrier housing (1) and cup for the right side bearing (4). To check preload adjustment using special tool (MF 245D or equivalent) refer to Fig. 217.

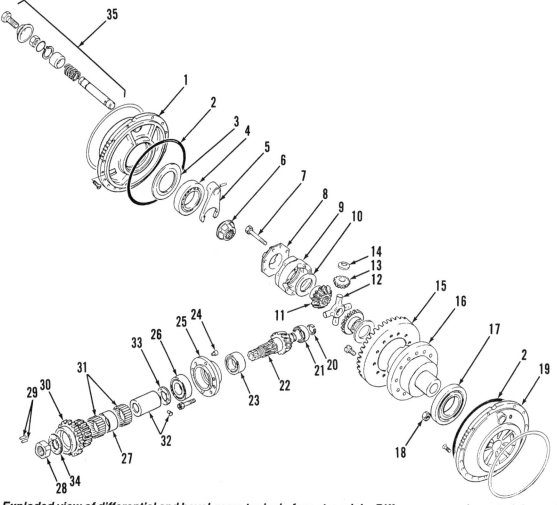

Fig. 216—Exploded view of differential and bevel gears typical of most models. Differences may be noted depending upon pto drive.

1. Carrier housing (Right)
2. "O" ring
3. Spacer shield
4. Bearing cup & cone
5. Differential lock fork
6. Jaw coupler
7. Bevel pinion gear
8. Coupler cap
9. Differential case half
10. Thrust washer
11. Side gear
12. Cross
13. Pinion gear
14. Thrust washer
15. Bevel ring gear
16. Differential case half
17. Bearing
19. Carrier housing (Left)
20. Snap ring
21. Bearing
22. Bevel pinion gear
23. Bearings
24. Locating pin
25. Retainer housing
26. Bearing
27. Spacer
28. Adjusting nut
29. Locking pins
30. Pto drive gear
31. Bearing
32. Sleeve & keeper
33. Thrust washer
34. Thrust washer
35. Differential lock shaft

Assemble differential assembly and left carrier housing (19—Fig. 216) in housing. Remove cup for right bearing (4) and the spacer shield (3) from the right carrier housing (1). Position the cup for bearing (4) over the bearing cone, then install bar (B—Fig. 217) of special tool over the bearing cup and cone as shown. Attach bar (B) to housing with two screws tightened to 27 N·m (20 ft.-lbs.) torque and install the two special blocks (T). Locate a straightedge (S) across blocks and measure gap between center of bar and straightedge with a feeler gauge as shown at (F). Install spacer shield (3—Fig. 216) equal to the clearance measured by feeler gauge (F—Fig. 217). Spacer shields may be marked for identification by a series of punch marks or dots to indicate thickness as follows:

Eight Marks . . . . . 0.45-0.53 mm (0.019-0.021 inch)
Seven Marks . . . . 0.61-0.66 mm (0.024-0.026 inch)
No Marks . . . . 0.74-0.79 mm (0.029-0.031 inch)
One Mark . . . . 0.86-0.91 mm (0.034-0.036 inch)
Two Marks . . . . . . 0.99-1.04 mm (0.039-0.041 inch)
Three Marks . . . . . 1.12-1.17 mm (0.044-0.046 inch)
Four Marks . . . . . 1.25-1.30 mm (0.049-0.051 inch)
Five Marks . . . . . . 1.37-1.42 mm (0.054-0.056 inch)

Install the spacer shield with curved section as shown in Fig. 218. Remainder of reassembly is reverse of disassembly procedure.

DIFFERENTIAL LOCK

All Models

178. OPERATION. The mechanically actuated differential lock assembly is standard equipment on all models. When the differential lock foot pedal is depressed, the axle half of coupler ((6—Fig. 216) is forced inward to contact the differential case half of coupler cap (8). If slippage is occurring at either wheel, depressing the lock pedal will lock the differential case to the right axle. The differential and both rear drive wheels will rotate together as a unit. Pressure between the coupler dogs should keep the differential lock engaged and foot pedal may be released. When ground traction on both rear wheels becomes equal again, coupler dog contact pressure will be relieved and the coupler will automatically disengage.

179. ADJUSTMENT. First, check operation of differential lock linkage. Make sure that return spring is installed, linkage parts are clean and spring in operating plunger is free to move. When disengaged, clearance at (12—Fig. 219), between operating lever and adjusting screw, should be 1.0 mm (0.040 inch). If incorrect, lift rubber boot, loosen locknut, then turn adjusting screw until clearance is correct. Recheck operation, then tighten locknut and reposition rubber boot if clearance is correct. Refer to Fig. 220 for cross section. Refer to paragraph 180 if operating mechanism is rusted, dirty or for any other reason can not be adjusted properly.

180. R&R AND OVERHAUL. To remove the differential lock, first drain transmission housing. Disconnect the brake and differential lock linkage, then refer to paragraph 183 for removing rear axle housing from right side. The operating shaft (5—Fig. 221) passes through right axle housing and the differential carrier housing (1). Clean dirt and rust from shaft (5) before attempting to remove.

If renewal is necessary, bushings for shafts (5) should be removed and installed using piloted drivers of the proper size. Make sure that plunger pin (24) is free to operate in bore, before assembling. Remainder of assembly is reverse of disassembly. Refer to paragraph 179 for adjustment procedure.

MAIN DRIVE BEVEL GEARS

All Models

181. REMOVE AND REINSTALL. The main drive bevel ring gear (15—Fig. 216) and bevel pinion gear (22) must be renewed as a matched set. Different reduction ratios are used in various applications. Normal installed reduction ratio of the bevel gears is as follows.

Standard Axle Models 362, 365, 375, 383, 390 and 390T

Ratio 4.375:1
Teeth . . . . 8 × 35

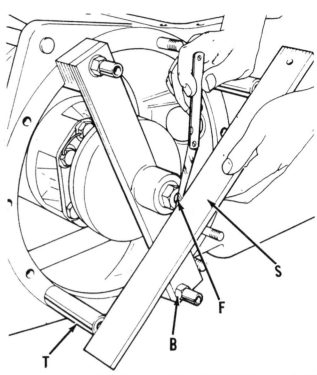

Fig. 217—View showing special tool MF245D, a straightedge (S) and feeler gauge (F) to determine correct thickness of spacer shield to install.

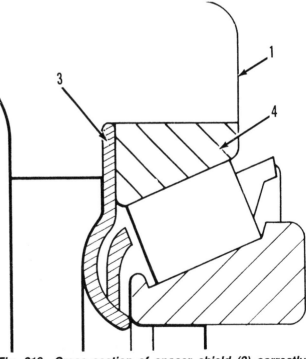

Fig. 218—Cross section of spacer shield (3) correctly installed. Refer to Fig. 216 for exploded view.

Standard Axle Models 390 and 398
    Ratio 3.889:1
    Teeth . . . . 9×35
Optional (40 km/hr) Axle Models
375, 390 and 398
    Ratio 3.454:1
    Teeth . . . . 11×38
All Models With 8 or 12 Speed Shuttle Gearbox
    Ratio 3.454:1
    Teeth . . . . 11×38

To remove the bevel ring gear, first remove the differential assembly as outlined in paragraph 176. The main drive bevel ring gear is secured to differential case half with either special bolts and nuts or with rivets. Rivets can be drilled out using a 13 mm (½ inch) drill after center punching rivet head. Retaining nuts are installed with thread locking compound and removal is easier if nuts are heated. The ring gear is a press fit on case and may also be epoxy bonded to case.

To remove the bevel pinion gear, first remove the differential as outlined in paragraph 176 and hydraulic lift cover as outlined in paragraph 218. Removal procedure is different depending upon pto and transmission type. Refer to the following procedures.

On models with independent pto, split tractor between differential housing and transmission housing as outlined in the appropriate paragraph in the transmission section. Remove pto output shaft and cover from left side of rear axle center housing. Disconnect hydraulic lines and remove pump support pins (S—Fig. 220A), then remove hydraulic pump and independent pto clutch unit through front of differential housing.

It is not necessary to split models not equipped with independent pto. It is, however, necessary to remove the coupler tube and pinion drive shaft. Remove side cover from left side of housing and move pto driven gear forward on pto shaft splines. Remove ground speed pto drive gear from pinion shaft.

On all models, remove screws attaching retainer (25—Fig. 216) to housing, then thread two of the screws into tapped, jackscrew holes in retainer. Tighten the two jackscrews evenly to force retainer and pinion assembly from housing bore.

Refer to paragraph 182 for overhauling the pinion and bearings assembly and to paragraph 176 for installing new ring gear. Refer to paragraph 177 to check and adjust differential carrier bearing preload, before installing the bevel pinion assembly.

Make sure that locating pin (24—Fig. 222) is in place and correctly aligned with hole in housing before pressing retainer (25) into place. Tighten screws attaching retainer (25) to 108 N·m (80 ft.-lbs.)

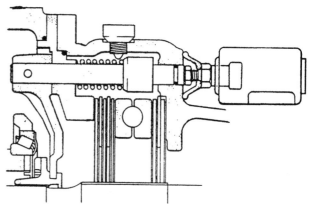

Fig. 220—Cross section of differential lock typical of all models.

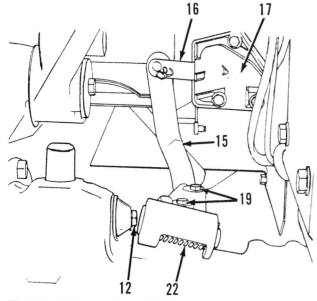

Fig. 219—External view of differential lock linkage typical of all models. Adjustment screw is shown at (12). Refer to Fig. 221 for legend.

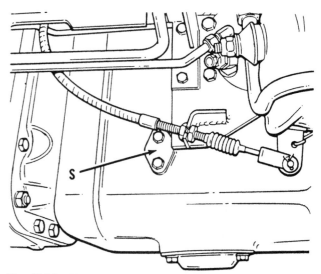

Fig. 220A—Hydraulic pump supports (S) are attached to each side of rear axle center housing.

torque. Complete assembly by reversing removal procedure. Always install new split pins, "O" rings and gaskets. Carefully locate the pump on ipto clutch and install supports (S—Fig. 220A). Tighten nuts or screws retaining support (S) to right side, but leave left side support retaining nuts or screws loose until tractor assembly is complete. Be sure that engine will not start and crank engine with starter, then tighten nuts or screws retaining support to left side of axle center housing. This procedure will ensure correct alignment of shafts.

182. OVERHAUL. Refer to paragraph 176 for renewal of bevel ring gear and for overhaul of the removed differential assembly. Refer to Fig. 222 for illustrations of typical bevel pinions and associated parts. Parts shown at top are typical of single speed pto, parts shown in middle view are typical of models with shiftable live or ground speed pto and parts shown in lower view are typical of two speed (540-1000 rpm) shiftable pto. Disassembly and reassembly procedures are similar for all models, even though

parts may be different. Refer to paragraph 181 for removal of pinion assembly.

Nut (28) is locked in place on all models by two locking pins (29) driven between shaft splines and threads of nut. To remove, it is necessary to destroy nut (28) by cutting. Make sure that new nut is available before removing. Clamp nut (28) in a soft jawed vice with flats of nut adjacent to locking pins (29) against jaws. Use a chisel to cut nut about ⅔ through at two places 90° from locking pins using a chisel. Nut should fracture during the second cut and make removal easy. Do not attempt to reuse nut. Bump pinion (22) from bearings after removing nut.

Examine all parts for wear or other damage and renew parts as necessary. Even though similar parts are identified by the same number in Fig. 216, parts may be different and the required parts may have different factory part numbers.

When reassembling, adjust preload of bearings (23 and 26) by tightening nut (28). Correct bearing preload will require 206-245 N·m (18-22 in.-lbs.) rolling torque to rotate pinion shaft in bearings. Bearing preload can be checked with a spring scale attached

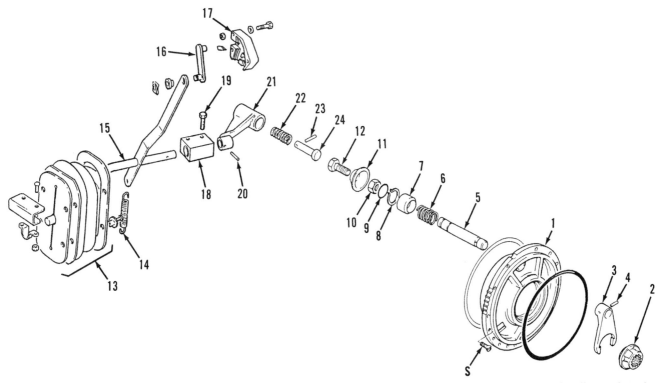

Fig. 221—Exploded view of differential lock mechanism typical of all models. Hydraulic pressure may be directed against shaft (5) of some models to move shaft and fork (3). Differential carrier housing (1) is attached to right side axle housing with screws (S).

1. Carrier housing	7. Collar	13. Control lever boot	19. Screws
2. Lock coupling	8. Snap ring	14. Return spring	20. Pin
3. Actuating fork	9. "O" ring	15. Control lever	21. Housing
4. Pin	10. Locknut	16. Bar	22. Spring
5. Shaft	11. Boot	17. Block	23. Plunger pin
6. Spring	12. Adjusting screw	18. Pivot	24. Operating pin

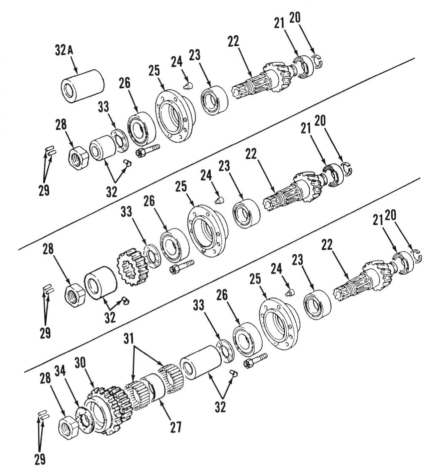

Fig. 222—Exploded view of typical bevel pinion gears and related parts. Parts shown at top are typical of single speed pto, parts shown in middle view are typical of models with shiftable live or ground speed pto and parts shown in lower view are typical of two speed (540-1000 rpm) shiftable pto. Sleeve (32A) for use with standard hydraulic pump is longer than similar sleeve for use with high volume pump.

20. Snap ring
21. Bearing
22. Bevel pinion gear
23. Bearings
24. Locating pin
25. Retainer housing
26. Bearing
27. Spacer
28. Adjusting nut
29. Locking pins (2 used)
30. Pto drive gear
31. Bearings
32. Sleeve & keeper
33. Thrust washer
34. Thrust washer

to string wrapped around pinion shaft and bearing retainer clamped in a vice. If rolling torque is checked with a spring scale and string, force should be 10-11 kg (21-25 lbs.). Lock adjustment by driving lock pins (29) into threads of nut in two adjacent splines under one of the flats. Drive locking pins in until flush with

front of locking nut. Recheck rolling torque after locking nut and tapping shaft to centralize bearings.

Lubricate pinion assembly thoroughly, align pin (24) with hole in housing, then tighten retaining nuts to 108 N·m (80 ft.-lbs.) torque.

REAR AXLE AND FINAL DRIVE

All tractors are equipped with a disc type brake located at the inner end of each axle and a planetary type reduction unit located in outer end of each rear axle housing.

REAR AXLE ASSEMBLY

All Models

183. REMOVE AND REINSTALL. Wheel axle shafts and planetary units can be removed without removing axle housings as outlined in paragraph 184.

To remove the complete axle shaft and housing as a unit, first drain oil from transmission housing, raise and support rear of tractor, then remove the rear

wheel. Release parking brake, then detach cable and return spring. Disconnect hydraulic brake lines and immediately cover all openings to prevent the entrance of dirt. If left axle housing is to be removed, remove necessary lines and the auxiliary hydraulic manifold. If right axle housing is being removed, remove linkage for differential lock that would interfere with removal and disconnect lock switch wiring. Unbolt fender or cab stabilizer bracket from outer end of axle housing and detach lower lift link from housing inner flange. Detach lift link from upper knuckle and remove the lower lift link. Remove the fender assembly or support the cab so that axle housing can be removed, then support the axle housing so that it can be separated from the tractor.

CAUTION: The axle housing assembly is heavy and awkward to handle. Be careful to support and move it safely. If axle housing is removed from left side, differential may fall from center housing. Support the differential unit as axle housing is removed to prevent it from falling.

Remove screws and nuts securing axle housing to the center housing, then carefully move the axle assembly away from center housing. Remove the "O" ring from carrier flange plate. If both axle assemblies are removed, install differential and left side axle housing before installing housing on right side.

Stick new "O" ring to center housing using petroleum jelly before positioning axle housing against center housing. Carefully align axle shaft splines with splines of differential side gears and holes in axle housing with attaching studs. Tighten retaining screws and stud nuts to 104-156 N·m (77-115 ft.-lbs.) torque. If necessary, refer to paragraph 177 to check and adjust differential carrier bearing preload. Remainder of assembly is the reverse of disassembly procedure. Coat threads of fender or cab stabilizer bracket with "Loctite 270" or equivalent, then tighten nuts to 230 N·m (170 ft.-lbs.) torque. Tighten wheel retaining nuts to 325 N·m (240 ft.-lbs.) torque. Fill transmission housing to correct level with oil and bleed air from brake lines.

PLANETARY ASSEMBLY

All Models

184. REMOVE AND REINSTALL. To remove the planetary assembly, first drain oil from planetary housing, raise and support rear of tractor, then remove the rear wheel. Lock parking brake ON and make sure that brake stays ON until assembly is completed to hold brake discs from falling. Scribe a mark across axle flange, ring gear (20—Fig. 223 or Fig. 224) and outer drive cover (6) to facilitate alignment when assembling, then remove nuts from screws holding planetary together. Lift outer housing, planetary assembly and ring gear away from axle housing, being careful not to withdraw inner axle from differential side gears and brakes.

Reinstall by reversing removal procedure. All sealing surfaces should be cleaned and inspected for damage that could cause leakage. Always install new gaskets on sides of ring gear. Tighten nuts securing outer housing and ring gear to 90-120 N·m (66-89 ft.-lbs.) torque on models with heavy duty axle. Tighten nuts securing outer housing and ring gear to 68-74 N·m (50-55 ft.-lbs.) torque on models with normal duty axles. The heavy duty axle with 4.8:1 reduction ratio is available on 390 and 398 models.

185. BEARING ADJUSTMENT. Planetary carrier bearing preload is adjusted by installing shims (25—Fig. 223 or Fig. 224) of different thicknesses under the bearing cup (24). Special tools (MF 267A, MF 267-1 and MF 267-2) are available for checking bearing preload and, if available should be used as follows:

Remove the wheel axle, planetary outer housing, planetary unit and ring gear as outlined in paragraph 184. Bolt the ring gear (20) to the outer housing (6) using four of the original bolts spaced evenly around circumference, using four of the wheel retaining nuts as spacers. Apply the handbrake fully and engage the differential lock (if right axle is being removed), then withdraw the drive axle (22) from the differential side gear, differential lock coupling and brake discs. It may be necessary to remove the complete axle hous-

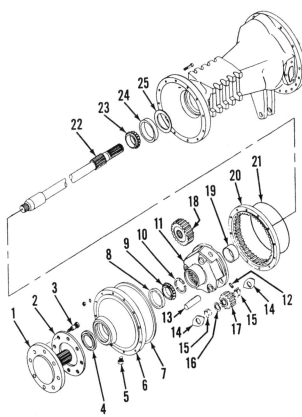

Fig. 223—Exploded view of normal duty wheel axle and planetary assembly used on 362, 365, 375, 383, 390 and 390T models. Heavy duty planetary, used on some models, is shown in Fig. 224.

1. Wheel axle	14. Thrust washer
2. Dust shield	15. Needle rollers
3. Wheel stud	16. Spacer washer
4. Seal	17. Pinion gear
5. Plug	18. Sun gear
6. Drive cover	19. Bushing
7. Gasket	20. Ring gear
8. Bearing cup	21. Gasket
9. Bearing cone	22. Drive axle
10. Thrust washer halves	23. Bearing cone
11. Planetary carrier	24. Bearing cup
12. Roll pin	25. Shim
13. Pinion shaft	

ing to reinstall drive axle if parts do not remain in alignment while drive axle is removed. Use a suitable puller to remove bearing cup (24) and any shims (25) that were previously installed. Use special sleeve (MF 267-1 for normal duty axle, MF 267-2 for heavy duty axle), position the removed bearing cup on the sleeve and install the assembled wheel axle and planet assembly into the bearing and special tool. Measure clearance between the ring gear and the two lugs of special tool. Clearance should be the same for each of the two lugs. Subtract the clearance measured from 1.0 mm (0.04 inch) and install shims (25) equal to this thickness in axle housing before installing bearing cone (24). Shims are available in thicknesses of 0.13, 0.25 and 0.38 mm (0.005, 0.010 and 0.015 inch). Refer to paragraph 184 for remainder of assembly.

If special tool is not available, remove the wheel axle, planetary outer housing, planetary unit and ring gear as outlined in paragraph 184. Apply the handbrake fully and engage the differential lock (if right axle is being removed), then withdraw the drive axle (22) from the differential side gear, differential lock coupling and brake discs. It may be necessary to remove the complete axle housing to reinstall drive axle if parts do not remain in alignment while drive axle is removed, but housing should remain on tractor while checking bearing preload. Use a suitable puller to remove bearing cup (24) and any shims (25) that were previously installed. Add approximately 0.762 mm (0.030 inch) thickness to the shims previously installed and reinstall bearing cup (24). Do not install drive axle (22), but install planetary assembly and ring gear using only three equally spaced retaining bolts finger tight. Bump outer end of wheel axle (1) sharply to make sure that bearings are seated. Equalize gap between axle housing and ring gear by tightening the three retaining bolts as required. When gap is equal, measure gap at three places next to the installed bolts and record the measurements. Remove the three retaining bolts, wheel axle, ring gear and planetary assembly. Remove the bearing cup (24) and remove shims equal to the gap plus 0-0.025 mm (0-0.0010 inch) to establish proper bearing preload. Reinstall drive axle and refer to paragraph 184 for remainder of assembly.

Normal Duty Axle and Planetary Unit

The 4.8:1, heavy duty planetary reduction unit available on 390 and 398 models is serviced using procedures described in paragraph 187. For servicing the normal duty unit with 3.75:1 installed on 362, 365, 375, 383, 390 and 390T models, refer to the following paragraph 186.

186. OVERHAUL. For servicing the normal duty unit with 3.75:1 reduction ratio, first refer to paragraph 184 and remove the planetary unit, then lift off

ring gear (20—Fig. 223). Drive out pinion shaft retaining pins (12), screw a 3/8 inch cap screw into threaded end of pinion shaft (13) and pull shafts from carrier. Carefully withdraw the three pinion gears (17) with thrust washers (14), loose rollers (15) and spacer washers (16). Each planetary pinion gear contains two rows of loose needle rollers (15) separated by a washer (16). Each path of rollers (15) contains twenty-nine individual rollers.

> **CAUTION: Do not lose, damage or mix rollers from other gears and make sure to account for all rollers, making sure that none stick to gear teeth or other surface.**

Slide sun gear (18) out through largest opening in planetary carrier (11). Drive bushing (19) into planetary carrier, then remove bushing. Remove inner bearing cone from planetary hub using an appropriate bearing puller, then use special tool (MF 265A or equivalent) and a 50 ton press to remove planet gear hub from axle splines. Remove both halves of thrust washer (10), then press wheel axle out of drive cover (6).

Clean, inspect and renew any components that are damaged or worn excessively. Petroleum jelly should be used to lubricate and hold components, including needle rollers, in position while assembling.

Dust shield (2) is retained by wheel studs and should not be removed unless renewal is necessary. Apply a light coat of "Hylomar" or similar sealer to outer surface of wheel axle seal (4), then install seal in drive cover (6) with metal side out and protruding 2 mm (0.080 inch) above flush with cover bore. Coat seal lip and fill cavity between seal and bearing cup (8) with petroleum jelly. Assemble wheel axle (1), drive cover (6) and bearing (8 and 9), pressing bearing cone (9) on shaft until fully seated.

Install thrust washer halves (10) and measure clearance between thrust washer halves and edge of shaft groove using a feeler gauge. Desired clearance is 0-0.025 mm (0-0.001 inch) and is adjusted by changing thickness of thrust washer halves which are available in thicknesses of 5.84-5.89, 5.90-5.94, 5.95-5.99, 6.00-6.04, 6.04-6.09, 6.10-6.14, 6.14-6.20, 6.20-6.25, 6.25-6.30 mm (0.230-0.232, 0.232-0.234, 0.234-0.236, 0.236-0.238, 0.238-0.240, 0.240-0.242, 0.242-0.244, 0.244-0.246 and 0.246-0.248 inch). Make sure that both halves are the same thickness and that thrust washer halves are fully seated in groove, then press the planet carrier (11) on wheel axle shaft until it bottoms.

Bushing (19) is presized and should not require machining after installation. Install bushing flush with chamfered edge in planet carrier bore. Preassemble planet gears (17), bearing rollers (15), spacer washer (16) and thrust washers (14), using petroleum jelly. Twenty-nine bearing rollers are in each path and each planet gear has two paths of bearings separated by a washer (16). Install sun gear (18) and the preassembled pinion gears in carrier (11). Larger thrust washers with squared ends are installed in the sun gear removal and installation opening. Insert pinion shafts (13) in carrier, aligning shaft retaining pin holes with corresponding holes in carrier, then install pins (12).

Install wheel axle and planetary assembly and adjust planetary carrier bearing preload as outlined in paragraphs 184 and 185.

Heavy Duty Axle and Planetary Unit

The normal duty planetary reduction unit used on 362, 365, 375, 383, 390 and 390T models is serviced using procedures described in paragraph 186. To service the 4.8:1 reduction ratio, heavy duty planetary reduction unit available on 390 and 398 models, refer to the following paragraph 187.

187. OVERHAUL. To service the heavy duty planetary reduction unit, first refer to paragraph 184 and remove the planetary unit, then lift off ring gear (20—Fig. 224). Drive out pinion shaft retaining pins (12), then pull pinion shafts (13) from planetary carrier (11). Carefully withdraw three pinion gears (17) with thrust washers (14), loose needle rollers (15) and spacer washers (16). Each planetary pinion gear contains two rows of loose needle rollers (15) separated by a washer (16). Each path of rollers (15) contains twenty-two individual rollers.

> CAUTION: Do not lose, damage or mix rollers from other gears and make sure to account for all rollers, making sure that none stick to gear teeth or other surface.

Slide sun gear (18) out through one of the openings in planetary carrier (11). Use special tool (MF 265A or equivalent) and a 50 ton press to remove the planetary carrier from the wheel axle shaft splines. Remove snap ring (10) from groove in axle shaft, then use a suitable press to remove wheel axle shaft (1) from the drive cover (6) and bearing cone (9).

Clean, inspect and renew any components that are damaged or worn excessively. Petroleum jelly should be used to lubricate and retain components while assembling.

Dust shield (2) is retained to wheel axle (1) by wheel studs and should not be removed unless renewal is necessary. Apply a light coat of "Hylomar" or similar sealer to outer surface of wheel axle seal (4), then install seal in drive cover (6) with metal side out. Seal should be installed flush with cover bore. Coat seal lip and fill cavity between seal and bearing cup (8) with petroleum jelly. Assemble wheel axle (1), drive cover (6) and bearing (8 and 9), pressing bearing cone (9) on shaft only far enough to permit installation of snap ring (10). After snap ring is installed, press end of axle shaft until bearing is firmly seated against snap ring. Make sure snap ring is firmly seated in shaft groove, then press planetary carrier (11) on wheel axle shaft until it bottoms. Preassemble planet gears (17), bearing rollers (15), spacer washer (16) and thrust washers (14), using petroleum jelly. Twenty-two bearing rollers are in each path and each planet gear has two paths of bearings separated a by washer (16). Install sun gear (18) and the preassembled pinion gears in carrier (11). Insert pinion shafts (13) in carrier, aligning shaft retaining pin holes with corresponding holes in carrier, then install pins (12).

Install wheel axle and planetary assembly and adjust planetary carrier bearing preload as outlined in paragraphs 184 and 185.

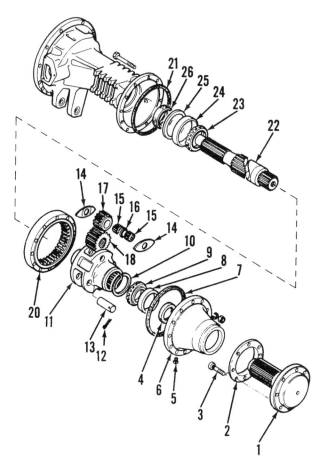

Fig. 224—Exploded view of heavy duty wheel axle and planetary assembly used on some 390 and 398 models.

1. Wheel axle	14. Thrust washer
2. Dust shield	15. Needle rollers
3. Wheel stud	16. Spacer washer
4. Seal	17. Pinion gear
5. Plug	18. Sun gear
6. Drive cover	20. Ring gear
7. Gasket	21. Gasket
8. Bearing cup	22. Drive axle
9. Bearing cone	23. Bearing cone
10. Snap ring	24. Bearing cup
11. Planetary carrier	25. Shim
12. Retaining pins	26. Seal
13. Pinion shaft	

DRIVE AXLE

All Models

188. REMOVE AND REINSTALL. To remove the drive axle (22—Fig. 223 or Fig. 224), first remove the planetary assembly as outlined in paragraph 184. Apply the handbrake fully and engage the differen-

tial lock (if right axle is being removed), then withdraw the drive axle (22) from the differential side gear, differential lock coupling and brake discs. It may be necessary to remove the complete axle housing, as outlined in paragraph 183, to reinstall drive axle if parts do not remain in alignment while drive axle is removed. Bearing cup (24) and shims (25) can be removed using a suitable puller.

BRAKES

All models are equipped with a planetary reduction unit at the outer ends of axles and disc brake units are located at axle inner ends, inside the axle housings. Brakes are operated hydraulically, but different types of slave cylinders and linkage are used. On tractors up to serial number V39466, the slave cylinder is shown in Fig. 228. The slave cylinder used on tractors after serial number V39465, is shown in Fig. 229.

ADJUSTMENT

Models Before Serial Number V39466

189. Raise and block rear of tractor so rear wheels are off the ground. Tighten adjusting nut (A—Fig. 225) until brake is locked, then loosen adjusting nut until wheel turns with only slightly perceptible drag. The nuts (A) should be backed off an equal amount for brake assemblies on both sides. Normally the adjusting nut of 362 models with three brake friction

discs should be backed off five flats. The adjusting nut of other tractor models with four friction discs should be backed off six flats. Check to make sure that brakes are adjusted equally on both sides.

Adjust parking brake cables by turning nuts (N) located on each side. Adjust cable until hand lever locks at about the third notch. Be sure that cables on both sides are adjusted evenly.

Models After Serial Number V39465

190. Raise and block rear of tractor so rear wheels are off the ground. Tighten adjusting nut (A—Fig. 226) until brake is locked, then loosen adjusting nut until wheel turns with only slightly perceptible drag. The nuts (A) should be backed off an equal amount for brake assemblies on both sides. Normally the adjusting nut of 362 models with three brake friction discs should be backed off five flats. The adjusting nut

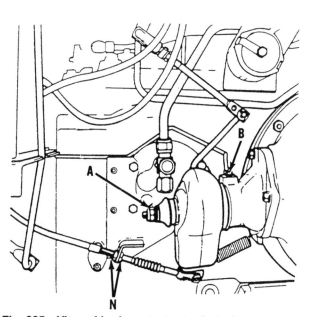

Fig. 225—View of brake actuator typical of type used on all models before serial number V39466.

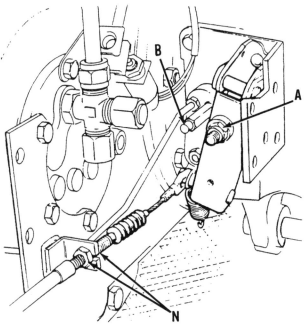

Fig. 226—View of brake actuator typical of type used on all models after serial number V39465.

of other models with four friction discs should be backed off six flats. Check to make sure that brakes are adjusted equally on both sides.

Adjust parking brake cables by turning nuts (N) located on each side. Adjust cable until hand lever locks at about the third notch. Be sure that cables on both sides are adjusted evenly.

FLUID AND BLEEDING

All Models

191. The brake fluid reservoir should remain filled with a mineral (petroleum) based oil such as Massey-Ferguson Powerpart brake fluid number 3405 389 M1. Fluid level should be maintained at level marked on reservoir and not allowed to run dry.

Whenever the brake system has been drained, refill with approved fluid to the proper level, then bleed air from the system. Bleeding can be accomplished in a conventional manner using a commercially available pressure or vacuum bleeding canister. Follow manufacturer's instructions if using a commercial bleeding system. Air can also be bled from the system manually.

If equipped with trailer brake system, bleed left side at trailer brake valve first. On late models (1991 and later) with high speed gearbox (identified by "S" prefix to serial number) both left and right brakes must be bled at the same time, because a tube connects the two master cylinders to balance the braking action.

Connect a clear plastic tube to air vent screw, bleeder valve (B—Fig. 225 and Fig. 226), and insert free end of tube in container with enough brake fluid to cover end of tube. Open bleeder valve and operate the brake pedal quickly, several times until reservoir is nearly empty then refill reservoir. Repeat cycle of operating the pedal and refilling reservoir until fluid expelled from tube is free of air bubbles, then close bleeder valve. It is usually necessary to cycle approximately ½ liter (1 pint) of brake fluid before all air is removed. Check for solid pedal feel and bleed again if sponginess is felt. Expelled fluid should not be reused if dirty, but clean fluid can be reused if allowed to sit overnight to remove air bubbles.

MASTER CYLINDERS

All Models

192. Either master cylinder can be removed from mounting bracket after disconnecting the operating rod clevis from pedal and hydraulic lines at connections on master cylinder. Refer to Fig. 227 and remove clevis, locknut (1) and boot (2). Remove snap ring (4), washer (5) and operating rod (3). It may be necessary to use compressed air to remove piston assembly (6 through 13). Disassembly of piston assembly is conventional, but use extreme care to prevent damage to piston (7) and seal (6). Use only clean, approved brake fluid to clean parts.

When assembling, lubricate all parts liberally. Install washer (12) with flare away from shoulder of valve rod (11). Refer to paragraph 191 for bleeding and to paragraph 189 or 190 for adjustment.

SLAVE CYLINDERS

Models Before Serial Number V39466

193. Refer to Fig. 225 for typical installation. To remove slave cylinder from left side, first remove

Fig. 227—Exploded view of master cylinder typical of all models.

1. Locknut
2. Boot
3. Actuating rod
4. Snap ring
5. Washer
6. Seal cup
7. Piston
8. Spring retainer
9. Spring
10. Retainer
11. Valve rod
12. Washer
13. Seal
14. Master cylinder body

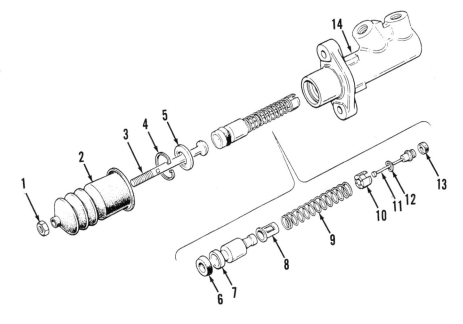

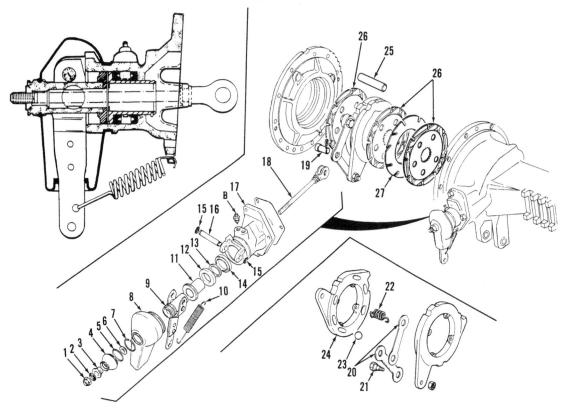

Fig. 228—Exploded view of slave cylinder typical of models before serial number V39466. The number of friction discs (26) and separator plates (27) may be different than shown. Cross section of actuator and slave cylinder is shown in inset.

1. Adjusting nut
2. Washer
3. Spacer
4. Boot
5. Spring ring
6. Washer
7. Snap ring
8. Boot
9. Handbrake lever
10. Return spring
11. Piston
12. Seal cup (same as 14)
13. Spacer
14. Seal cup (same as 12)
15. Snap ring
16. Pivot pin
17. Housing
18. Operating rod
19. Link pin
20. Links
21. Anchor bolts
22. Return springs (4 used)
23. Cam balls (6 used)
24. Expander plate (2 used)
25. Brake stop rod
26. Friction discs
27. Steel separator plate

pipes, fittings, hydraulic manifold and other interfering equipment as necessary. Remove the differential lock mechanism if removing right side unit. To remove unit from either side, remove adjusting nut (A—Fig. 225 or 1—Fig. 228), detach hydraulic brake line and cover openings to prevent the entrance of dirt. Disconnect the handbrake cable and remove the mounting screws. Disassembly is obvious after removing snap ring (7—Fig. 228), snap ring (15) and pivot pin (16). Use only clean, approved brake fluid to clean parts.

When assembling, lubricate all parts liberally. Refer to paragraph 191 for bleeding and to paragraph 189 for adjustment. Refer to paragraph 179 for adjustment of differential lock.

Models After Serial Number V39465

194. Refer to Fig. 226 for typical illustration. Slave cylinder and actuator can be removed as follows: To remove unit from left side, first remove pipes, fittings, hydraulic manifold and other interfering equipment as necessary. Remove the differential lock mechanism if removing right side unit. To remove unit from either side, detach hydraulic brake line, then screw plug into opening to seal opening and stop cylinder pistons from expanding and falling out while removing. A second bleeder plug can be used. Disconnect the handbrake cable, remove adjuster nut (1—Fig. 229), remove the three screws attaching actuator to axle housing, then withdraw unit from axle. Be careful not to damage boot or threads of rod (18).

Spring (13) will probably push at least one of the pistons (11) from bore when plug covering opening for the hydraulic line is removed. If necessary, air can be used to blow piston from bore. Make sure bleeder valve (B) is open, clean and operates freely in threads before installing. Make sure that threads of adjuster rod (18) are clean. Seals (12), gasket (19), boot (4), "O" ring (14), self-locking adjuster nut (1) and any questionable parts should be renewed when unit is disassembled.

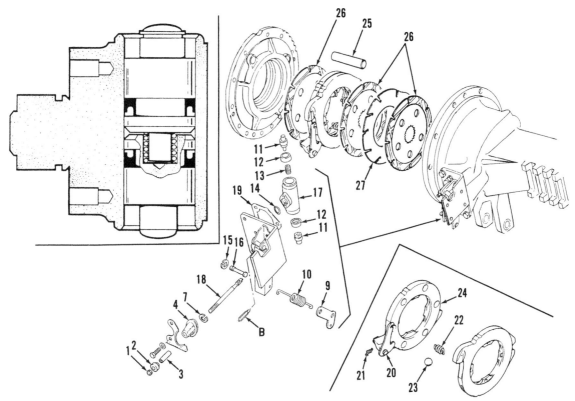

Fig. 229—Exploded view of actuator, slave cylinder and brake discs typical of models after serial number V39465. The number of friction discs (26) and separator plates (27) may be different than shown. Cross section of slave cylinder is shown in inset.

1. Adjusting nut	10. Return spring	16. Pivot pin	22. Return springs (4 used)
2. Bearing	11. Piston	17. Slave cylinder housing	23. Cam balls (6 used)
3. Spacer tube	12. Seal cup	18. Operating rod	24. Expander plate (2 used)
4. Boot	13. Spring	19. Gasket	25. Brake stop rod
7. Snap ring	14. "O" ring	20. Bellcrank	26. Friction discs
9. Handbrake lever	15. Snap ring	21. Bellcrank return spring	27. Steel separator plate

Lubricate all parts liberally while assembling. Remove seals (12) from grooves of pistons (11), being careful not to leave any marks on piston. Install new seals (12) in groove with cup side toward center of cylinder. Assemble pistons and seals (11 and 12) and the spring (13) in cylinder, hold the pistons together and install bleeder valves (B) in both the bleeder port and the line connection. With both ports closed, pistons should remain together compressing spring to facilitate remainder of assembly. Coat new "O" ring (14) and boot (4) with "Hylomar" or equivalent sealer before installing. Assemble cylinder to actuator plate and tighten retaining screws to 23-29 N·m (17-21 ft.-lbs.) torque. If removed, reattach adjuster rod (18) to bellcrank (20) and make sure that spring (21) is operating properly.

When reinstalling, make sure that sealing surfaces of axle housing and actuator plate are clean and all of gasket (19) is removed. Coat gasket surface of axle housing with "Loctite 518" or equivalent and slide protector sleeve (MF 467 or equivalent) over threads of adjuster rod (18). Slide hand brake actuator and brake slave cylinder into position, being careful not

to damage boot (4) and carefully guiding cylinder and pistons (11) into place between lugs of expander plates (24). Tighten the three retaining screws to 12-16 N·m (9-12 ft.-lbs.) torque. Install adjuster nut (1) and bearing (2), then reattach hand brake cable. Remove plug and reattach hydraulic line, then bleed air from system as outlined in paragraph 191. Refer to paragraph 190 for adjustment. Refer to paragraph 179 for adjustment of differential lock.

DISC BRAKE ASSEMBLY

All Models

195. Refer to paragraph 183 for removing rear axle housing. Refer to paragraph 193 or paragraph 194 and remove the brake slave cylinder and actuator unit shown in Fig. 228 or Fig. 229. Refer to paragraph 180 for removal of differential lock assembly. Refer to Fig. 221 and remove screws (S) that attach the differential carrier housing (1) to the axle housing. Remove carrier housing and refer to Fig. 228 or Fig. 229.

On early models, remove pin (19—Fig. 228) and operating rod (18). Remove brake stop rod (25) and discs (26), plates (27) and expander assembly (20, 21, 22, 23 and 24). Model 362 tractors are equipped with three friction discs (26) and one steel plate (27) assembled as shown. Other models are equipped with four friction discs (26) and two steel plates equally divided between sides of expander assembly. Depth of groove in friction disc should be more than 0.3 mm (0.012 inch).

On later models, detach rod (18—Fig. 229) from bellcrank (20). Remove brake stop rod (25) and discs (26), plates (27) and expander assembly (20, 21, 22, 23 and 24). Model 362 tractors are equipped with three friction discs (26) and one steel plate. Two friction discs and the one steel plate are located between wheel and expander. Other models are equipped with four friction discs (26) and two steel plates equally divided between sides of expander assembly. Depth of groove in friction disc should be more than 0.3 mm (0.012 inch).

Reassembly is reverse of disassembly. Make sure that friction discs and steel plates are installed in order described. Refer to paragraph 180 for installation of differential lock assembly and to paragraph 193 or paragraph 194 for installation of the brake slave cylinder and actuator unit. Refer to paragraph 183 for installing axle housing. Refer to paragraph 191 and bleed brakes, then refer to paragraph 189 or 190 to adjust brakes.

LIVE POWER TAKE-OFF

196. All models are equipped with either a dual or split torque clutch. The dual clutch controls engagement of drive to the engine transmission and to the live power take off (pto). Split torque clutches are installed on tractors equipped with independent pto. Refer to the appropriate following paragraphs for service to live pto.

OUTPUT SHAFT, SEAL AND REAR BEARING

Models With Live Pto

197. To remove pto output shaft, drain oil from axle center housing and transmission, shift pto control to engaged live position and remove pto shield. Detach bottom of 3-point hitch center control beam, remove retainer attaching nuts (2 and 4—Fig. 230), then remove the two center screws. Unbolt and remove retainer (3) and plate (5), then use threaded cap (1) to pull seal retainer (7) from center housing. Shaft (9) will probably be removed with seal and rear bearing. Press bearing (10) onto shaft and seal into housing (7). Rear of seal (6) should be 1 mm (0.04 inch) below edge of housing. Renew "O" ring (8), coat seal (6) and "O" ring liberally with transmission oil and install shaft bearing and seals into center housing. Drive seal housing into center housing and install retainer (3).

Reverse removal procedure to install output shaft and seal. Tighten the four nuts with spacers (4) to 113-169 N·m (83-125 ft.-lbs.) torque and the two nuts (2) attaching drawbar to chain bracket to 230-260 N·m (170-192 ft.-lbs.) torque.

Output shaft front needle bearings can be renewed after first removing the differential assembly as outlined in paragraph 176. Bearing should be removed toward the rear. Use an appropriate size driver to

Fig. 230—View of pto output shaft and associated parts for models with live pto.

1. Cap
2. Nuts to drawbar bracket
3. Retainer
4. Nuts, washers and spacers
5. Seal housing plate
6. Triple lip seal
7. Seal housing
8. "O" ring
9. Output shaft
10. Rear bearing
11. Snap ring

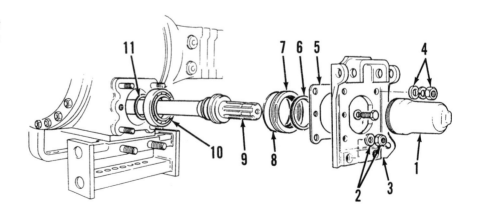

install new needle bearing just below flush with bore at rear.

ENGAGEMENT COUPLING AND GROUND SPEED GEARS

Models With Live Pto

198. To remove engagement coupling, two speed gears or ground speed gears, remove rear shaft as outlined in paragraph 197, then refer to paragraph 181 for access to the coupling and, if so equipped, the ground speed drive gears.

PTO FRONT DRIVE SHAFTS AND GEARS

Models With Live Pto

199. Refer to the appropriate transmission section for removing or servicing pto drive shafts, gears or bearings which transfer drive from the engine clutch to the pto engagement coupling.

INDEPENDENT POWER TAKE-OFF

200. All models are equipped with either a dual or split torque clutch. Split torque clutches are installed on tractors equipped with independent pto. Drive for the hydraulic pump and for the pto are taken from the splined hub of a plate attached to the clutch cover, which turns all the time that engine is running. Control of pto engagement or disengagement is by a separate, hydraulically operated clutch. The dual clutch controls engagement of drive to the engine transmission and to the live power take off (pto). Refer to the appropriate following paragraphs for service to independent pto.

hub of clutch cover and turns continuously when engine is running.

The pto control lever is on the left side cover of differential housing. Moving lever forward disengages the hydraulically actuated multiple disc clutch and engages the pto brake. Moving lever rearward releases the hydraulic brake and engages the multiple disc pto clutch.

Hydraulic pressure to actuate the pto clutch and brake is produced by the auxiliary gear type pump which also provides power for the Multi-Power clutch and/or auxiliary hydraulic system if so equipped.

OPERATION

Models With Independent Pto

201. Some tractors may be equipped with an independent pto which is driven by a flywheel "Split Torque" clutch and controlled by a hydraulically actuated multiple disc clutch contained in the differential housing. The pto front drive shaft is splined into

OUTPUT SHAFT

Models With Independent Pto

202. REMOVE AND REINSTALL. To remove the output shaft (9—Fig. 231), first drain oil from axle center housing and transmission and remove pto shield. Detach bottom of 3-point hitch center control beam, remove retainer attaching nuts (2 and 4), then remove the two center screws. Unbolt and remove

Fig. 231—View of pto output shaft and associated parts for models with independent pto.

1. Cap
2. Nuts to drawbar bracket
3. Retainer
4. Nuts, washers and spacers
5. Seal housing plate
6. Triple lip seal
7. Seal housing
8. "O" ring
9. Output shaft
10. Rear bearing
11. Snap ring
12. Bearing sleeve
13. "O" ring
14. Snap ring

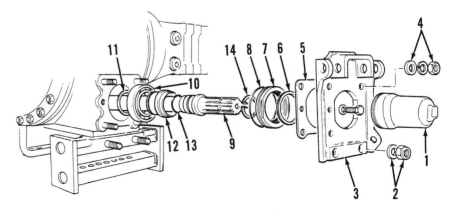

retainer (3) and plate (5), then use threaded cap (1) to pull seal retainer (7) from center housing. Shaft (9) will probably be removed with seal and rear bearing.

Press bearing (10) onto shaft and seal into housing (7). Rear of seal (6) should be 1 mm (0.04 inch) below edge of housing. Renew "O" ring (8), coat seal (6) and "O" ring liberally with transmission oil and install shaft bearing and seals into center housing. Drive seal housing into center housing and install retainer (3).

Reverse removal procedure to install output shaft and seal. Tighten the four nuts (4) with spacers to 113-169 N·m (83-125 ft.-lbs.) torque and the two nuts (2) attaching drawbar to chain bracket to 230-260 N·m (170-192 ft.-lbs.) torque.

Output shaft front needle bearings can be renewed after first removing the differential assembly as outlined in paragraph 176. Bearing should be removed toward the rear. Use an appropriate size driver to install new needle bearing just below flush with bore at rear.

LEFT SIDE COVER AND CONTROL VALVE

Models With Independent Pto

203. REMOVE AND REINSTALL. Drain transmission until level of oil is at low mark on dipstick. Detach control rods and hydraulic lines that would interfere with removal of side cover and remove or relocate as necessary. Remove plug (1—Fig. 232) and locator plug (2) and nuts (3 and 4). Unbolt and remove cover, making sure that tubes do not catch.

When installing, install new gasket, "O" rings and sealing washers. Be sure that internal lever engages the control valve. Coat gasket face of side cover with "Loctite 515 Instant Gasket" or equivalent. Reassemble by reversing removal procedure. Seal the six screws retaining side cover with "Hylomar" or equivalent sealer.

PTO CLUTCH, BRAKE AND CONTROL VALVE

Models With Independent Pto

204. REMOVE AND REINSTALL. To remove the clutch and control valve unit, first remove hydraulic lift cover as outlined in paragraph 218 and split tractor between differential housing and transmission housing as outlined in the appropriate paragraph in the transmission section. Remove pto left side cover, disconnect hydraulic lines and remove pump support pins, then remove hydraulic pump and independent pto clutch unit through front of differential housing.

To reinstall clutch assembly, reverse the removal procedure.

205. OVERHAUL. Refer to Fig. 233 for an exploded view of clutch, brake and control valve assembly. To disassemble the removed clutch, first remove snap ring (20), thrust washer (19) and control valve housing (18) from clutch hub end of housing (15). Remove retaining ring (4) and withdraw retainer plate (1), thrust washer (2), drive hub (3), shims (5), wave springs (6), separator plates (7) and friction plates (8) from clutch housing. Tap clutch housing against wooden bench to remove piston front plate (10), Belleville washer (11), spacer (13) and clutch piston (14).

To disassemble control valve, remove detent plug, spring and ball (21). Drive out roll pin (23), then withdraw plunger (33) and valve spool (27) assembly from valve body. Unseat snap ring (32) and separate valve spool from plunger. Drive roll pin (29) from spring guide (30) and remove guide and spring (28) from valve spool. Remove retaining ring (24), then tap sleeve (26) with "O" ring from valve body bore.

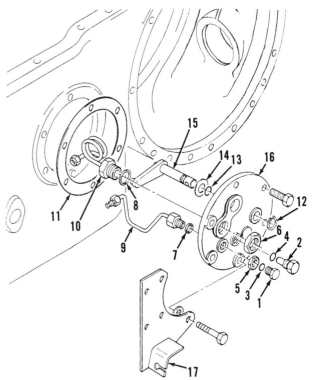

Fig. 232—Exploded view of left side cover showing hydraulic lines and valve control lever.

1. Plug	10. Tube
2. Locator plug	11. Gasket
3. "O" ring	12. Snap ring
4. "O" ring	13. "O" ring
5. Nut	14. Washer
6. Nut	15. Control valve
7. "O" ring	operating lever
8. Washer	16. Left side cover
9. Tube	17. Plate

Unbolt and remove brake pad (40) and brake cylinder (37) from control valve housing. Remove retaining ring (34) and withdraw piston (39) from cylinder.

Renew all "O" rings and sealing rings. Renew clutch separator rings and friction discs if excessively worn, warped or discolored from overheating. Free height of wave springs (6) should be 5.21 mm (0.205 inch) and all seven should be within 0.51 mm (0.020 inch) of the same height. If new friction discs are installed, soak discs in transmission oil for at least 30 minutes before assembling.

When assembling pto brake, a new shoulder bolt (41) should be installed to retain brake pad. Tighten shoulder bolt and other screw to 7 N.m (60 inch lbs.) torque. Be sure that brake pad moves freely.

To reassemble control valve, reverse the disassembly procedure. Insert roll pin (23) with split side of pin facing away from plunger (33). Tighten detent plug to 47 N.m (35 ft.-lbs.) torque. Piston rings (17) are installed in grooves of housing (15), but should have end gap of 0.050-0.305 mm (0.002-0.012 inch) in bore of housing (18).

Lubricate piston seal rings with clean transmission/hydraulic oil when assembling. Install piston, flat side first, and spacer ring into clutch housing. Install Belleville washer with concave (dished) side outward, then install piston front plate with stepped edge inward. Install the seven steel plates (7), seven wave springs (6) and seven friction discs (8) alternately, beginning with an externally splined steel separator plate (7), followed by a wave spring (6) and finally a friction disc (8). New friction discs should be soaked in oil before assembling. To facilitate assembly, small pins can be inserted through the three vent holes in clutch housing (15) to hold plates and springs compressed. Remove pins before completing assembly.

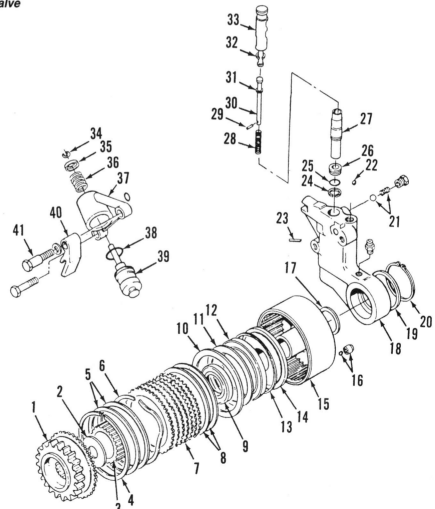

Fig. 233—Exploded view of typical ipto multiple disc clutch, control valve assembly and brake assembly.

1. Retainer and gear
2. Thrust washer
3. Drive hub
4. Retaining ring
5. Shims
6. Wave springs (7 used)
7. Separator plates (7 used)
8. Friction discs (7 used)
9. Piston seal
10. Piston front plate
11. Belleville washer
12. Piston seal
13. Spacer
14. Clutch piston
15. Clutch housing
16. Check valve assembly
17. Seal rings (2 used)
18. Control valve housing
19. Thrust washer
20. Snap ring
21. Detent spring & ball
22. Ball
23. Roll pin
24. Retaining ring
25. "O" ring
26. Sleeve
27. Valve spool
28. Spring
29. Pin
30. Spring guide
31. Washer
32. Retaining ring
33. Plunger
34. Retaining ring
35. Spring seat
36. Return spring
37. Brake cylinder
38. "O" ring
39. Piston
40. Brake pad
41. Shoulder bolt

Select proper thickness of shims (5) as follows. Assemble retainer plate (1) over the last sintered friction disc (8), leaving out shims (5) and thrust washer (2), then install snap ring (4). Press retainer (1) into housing (15) with a force of about 64 kg. (140 lbs.) to compress wave springs, then measure distance between retainer (1) and snap ring (4). Gap should be 2.54-3.05 mm (0.100-0.120 inch). If gap exceeds 3.05 mm (0.120 inch), install 0.20-0.51 mm (0.008-0.020 inch) or 0.30-0.76 mm (0.012-0.030 inch) thick shims as necessary. If available, an extra retainer plate (1) can be modified by machining gear away to provide more room for measuring clearance. **The modified retainer should be used only for measuring and should not be installed.** Remove snap ring (4) and retainer plate (1), then install thrust washer (2) and selected shims (5). Reinstall retainer and be sure that snap ring is fully seated in groove.

HYDRAULIC PRESSURE TESTS

Models With Independent Pto

206. Remove test plug (1—Fig. 232) from left side cover and attach a suitable 30 bar (400 psi) pressure test gauge to the port. Start engine and run until transmission oil reaches normal temperature. Set engine speed to 1200 rpm, engage pto and observe pressure indicated at test gauge. Pressure should be 15.5-19 bar (225-275 psi) for models with single speed pto.

HYDRAULIC SYSTEM

207. All models are equipped with basically two separate hydraulic systems: A main hydraulic system and an auxiliary hydraulic system. Additionally, the auxiliary system is divided into two separate systems, one supplying pressure for steering, ipto clutch and brake and Multi-Power clutch, the second supplying pressure for external cylinders and spool valves.

The main hydraulic system consists of a four-piston type pump submerged in hydraulic fluid in the differential housing and driven by the pto drive shaft. A control valve is located in the pump unit which meters the operating fluid at pump inlet. A single-acting hydraulic cylinder, attached to bottom of lift cover, actuates the hitch rockshaft.

The two-section, gear type, auxiliary pump is installed on the right side of the engine and driven by the engine gears or auxiliary drive shaft. Oil is filtered through a 140 micron inline filter before entering either of the two pump sections. One pump section provides pressure for steering, ipto and Multi-Power clutch. The second pump section delivers pressurized oil to the auxiliary hydraulics.

Refer to the appropriate following paragraphs for service to main hydraulic system or to paragraphs 226 through 230 for service to the auxiliary hydraulic system.

RESERVOIR AND FILTERS

All Models

208. The transmission, transfer drive (four-wheel-drive models), differential, rear axle, pto and hydraulic systems share the same fluid. Transmission housing, center housing and rear axle housings serve as system reservoir and are each provided with individual drain plugs. Recommended fluid is "Massey-Ferguson Super 500" multi-use 10W-30 oil, "Massey-Ferguson Permatran" or equivalent oil. Capacity is approximately 43.4 L (11.5 gal.) **without** spacer or four wheel drive or 47.4 L (12.5 gal.) **with** spacer or four-wheel drive. Specific options may add or reduce the amount of oil and, if drained, varying amounts of oil will cling to internal parts and walls of housings. Maintain oil at full level of dipstick shown in Fig. 234 or Fig. 235.

The main hydraulic system inlet strainer, located in bottom of rear axle center housing, should be removed and cleaned or renewed every 1000 of operation. Drain all oil from the transmission and rear axle center housings, then remove the three screws retaining cover (1—Fig. 236) to bottom of center housing. Remove clip (3), nut (4), filter (8) and remaining parts from opening. Clean all parts thoroughly and install new gasket and "O" rings. Reinstall inlet filter assembly and drain plugs, then refill with clean approved oil.

Oil entering the two section auxiliary pump is screened by filter (4—Fig. 237) located in the suction line. This filter should be removed, cleaned or renewed and reinstalled every 1000 hours of operation. Drain approximately 10 L (2 gal.) of oil from transmission housing, before unbolting and removing cover (1). Install new gaskets, then tighten retaining bolts evenly and securely.

Pressurized oil not required for operating the power steering system, ipto clutch and Multi-Power clutch is circulated through a cooler and filter (F—Fig. 238) before being directed to the auxiliary mani-

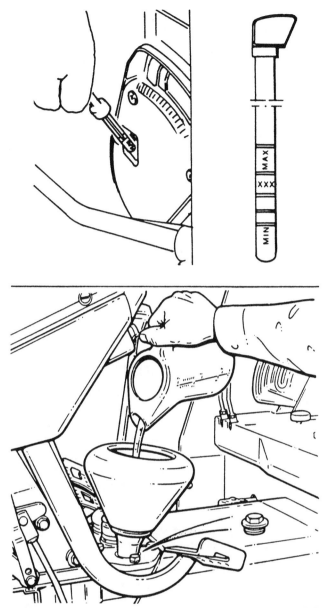

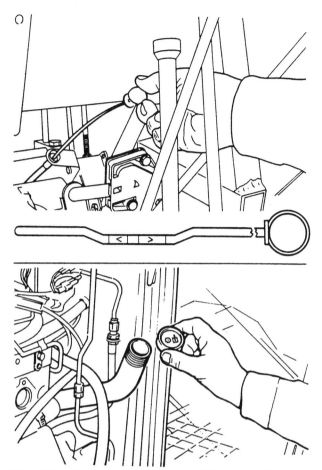

Fig. 235—Transmission oil level of tractors with cab is measured by dipstick that points toward rear as shown.

Fig. 234—Transmission oil level of tractors without cab is measured by dipstick located in the side cover as shown. Refer to Fig. 235 for models with cab.

fold. This canister type filter should be removed and new filter installed after each 250 hours of operation. Wipe seal at top of canister with oil before installing, then run engine and check for leaks.

TROUBLESHOOTING

All Models

209. The following is a list of symptoms which may occur during operation and their possible causes. These suggestions can be used in conjunction with other Tests and Adjustments to isolate the cause of problems before proceeding with component disassembly.

1. Hitch will not raise. Could be caused by:
   a. Internal oil leak in system. Remove response control side cover and check for leakage.
   b. Damaged, binding or maladjusted control valve linkage.
   c. Maladjusted or faulty servo valve (pressure control units) or faulty safety relief valve (non-pressure control units).
   d. Faulty main hydraulic pump.

2. Lift links do not raise evenly. Jerky operation could be caused by:
   a. Valve sticking in pump valve chamber.

3. Lift links will not raise to full height. Could be caused by:
   a. Transport stop maladjusted.
   b. Control valve maladjusted.
   c. Control linkage maladjusted.

4. Lift links will not lower. Could be caused by:
   a. Control valve sticking.
   b. Control valve maladjusted.
   c. Lift arms binding.

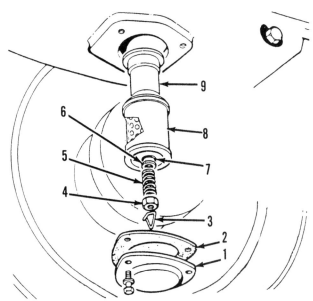

Fig. 236—Main hydraulic system inlet strainer should be removed and cleaned or renewed every 1000 hours of operation. A large "O" ring, not shown, is located above tube (9).

1. Cover
2. Gasket
3. Lock clip
4. Nut
5. Spring
6. Washer
7. "O" ring
8. Filter
9. Tube

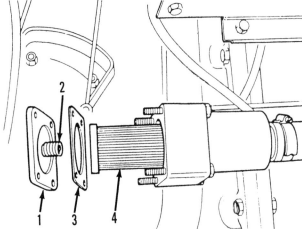

Fig. 237—Inlet screen shown is located on inlet line to auxiliary pump and filters oil entering both steering and auxiliary sections of pump.

1. Cover
2. Spring
3. Gasket
4. Filter

5. Hitch lowers too fast with response control in "Slow" position. Could be caused by:
  a. Response control maladjusted.
  b. Control linkage binding or damaged.
6. Erratic action when operating in draft control. Could be caused by:
  a. End play in master control spring.

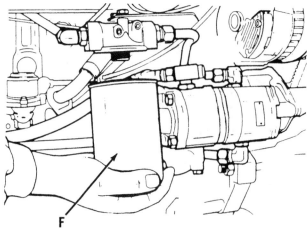

Fig. 238—Canister filter (F) should be renewed every 250 hours of operation.

  b. Control linkage binding or damaged.
7. Lift links creep down. Could be caused by:
  a. Internal leakage in lift cylinder, control valve or valve chambers.

TESTS AND ADJUSTMENTS

All Models

210. All operational tests should be conducted with hydraulic fluid at normal operating temperature of about 50-70° C (120-160° F). Always cover open lines and ports as soon as possible to prevent the entrance of dirt and unnecessary fluid loss.

> **CAUTION: To prevent injury, engine should be stopped, all hydraulically actuated equipment should be rested on the ground and all hydraulic pressure within circuits relieved before removing any test port or disconnecting any hydraulic line.**

211. PRESSURE. Remove test plug from port (P—Fig. 239) on left side of lift cover and connect a suitable 300 bar (4000 psi) pressure test gauge to the port. If equipped with selector valve, move control lever to "Linkage" position. Start engine and operate at 1200 rpm, move the Position Control lever to "Constant Pumping" position and observe pressure indicated by test gauge. Correct pressure is 221-231 bar (3200-3350 psi) for all models. Some models (from serial number N42314, manufactured since November, 1988) are equipped with a non-adjustable cartridge, relief valve setting is adjustable on other models. Relief valve is accessible through opening after removing cover from right side of rear axle center housing.

212. MASTER CONTROL SPRING. Move the draft control lever to "Down" position. Disconnect control beam from spring head (1—Fig. 239) and

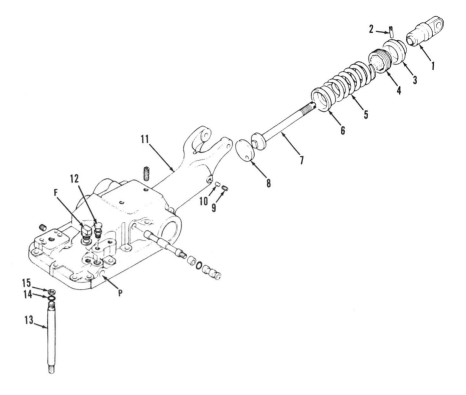

Fig. 239—Exploded view of master control spring assembly used on all models. Main pump pressure may be tested at port (P).

F. Filler opening
P. Test port
1. Head
2. Pin
3. Boot
4. Adjusting nut
5. Control spring
6. Spring guide
7. Plunger
8. Overload stop disc
9. Set screw
10. Nylon plug
11. Lift cover
12. Breather
13. Standpipe
14. "O" ring
15. Washer

pivot beam down to expose spring head. Check for end play by pulling and pushing spring head. If end play is indicated, remove Allen set screw (9) from side of housing and pull boot (3) back to expose adjusting nut (4). Unscrew nut (4) from lift cover using special spanner wrench and withdraw spring assembly from top cover. Nylon plug (10) will be damaged and should be removed so that new plug can be installed.

Hold spring head (1) and attempt to turn spring (5). Spring should fit snugly with no end play, but should still turn with moderate effort. If adjustment is incorrect, drive pin (2) out and thread spring head (1) onto or from plunger (7) until end play is just eliminated. Tighten plunger as required until slot is aligned and reinstall pin (2). Reinstall control spring assembly in lift housing bore and tighten nut (4) until end play is just eliminated. If adjusting nut is either too loose or too tight, unit will have noticeable end play. Install new nylon plug (10) and tighten Allen set screw (9) to 7 N·m (60 inch lbs.) torque when adjustment is correct.

213. EXTERNAL LINKAGE ADJUSTMENTS.
The following adjustments should be performed with the lift cover installed and the tractor operational. Loosen quadrant bolts (1—Fig. 240), locate quadrant in center of slotted hole, then tighten bolts. Move the Draft Control lever (2) to between "Sector" marks and Position Control lever (3) to the "Transport" position. Locating holes (5) for position control and locating holes (6) for draft control should both be aligned so that 6.5 mm (¼ inch) diameter locating pins can be inserted. Linkage attaching holes (N) are used if tractor is not equipped with spacer or four-wheel-

drive transfer case. Linkage is attached to holes (S) if spacer is located between transmission and rear axle center housings. Remove pins from holes (5 and 6) after adjusting lengths of control rods (7) if required.

Attach a weight of at least 400 kg (900 lbs.) to the lower lift links, start engine and set engine speed to 1200 rpm. Top link should not be attached. Cycle controls throughout operating range at least six times to bleed air from system. Move the Position Control lever (3) to "Transport" and the Draft Control lever (2) between "Sector" marks. Loosen quadrant mounting bolts (1) and move the draft control quadrant and lever until the lower links are held in horizontal position. Tighten quadrant mounting bolts (1) and recheck alignment of holes (5 and 6) as described in previous paragraph.

Move Draft Control lever to the fully UP position and the Position Control lever to "Transport." Loosen quadrant mounting bolts (1) and rotate the position control quadrant and lever until the rear linkage reaches maximum travel (indicated by relief valve opening). Scribe a mark (A—Fig. 241) on lift cover and a similar line (B) on lift arm. Scribe a second mark (C) on lift cover 6 mm (¼ inch) to the rear and above the first mark. Rotate the position control quadrant and lever until the lift arms lower slightly and mark (B) on lift arm aligns with mark (C) on cover. Tighten bolts (1—Fig. 240) and recheck alignment of holes (5 and 6) as described in earlier paragraph.

Move the response lever (R—Fig. 242) so that it is 6 mm (¼ inch) from slow position as shown at (D).

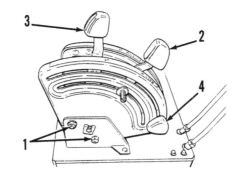
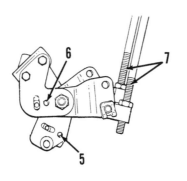

Fig. 240—Views of Position Control lever, Draft Control lever, related linkage and adjustment points. Refer to text for adjustment procedure.

1. Quadrant mounting bolts
2. Draft Control lever
3. Position Control lever
4. Adjustable stop
5. Locating hole for Position Control
6. Locating hole for Draft Control
7. Control rods

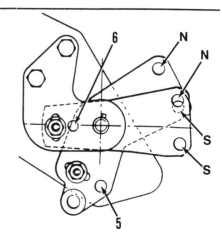

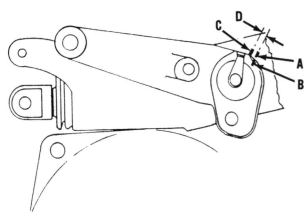

Fig. 241—Marks should be scribed on lift arms and lift cover as described in text to measure movement of lift arms when adjusting linkage. Distance (D) between marks on cover should be 6.5 mm (1/4 inch).

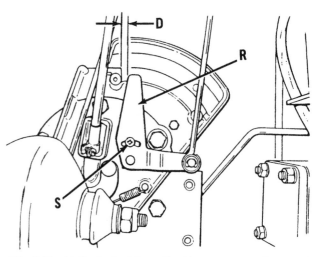

Fig. 242—Refer to text for adjusting response lever (R).

Loosen screw (S), rotate inner cam clockwise by moving the locking screw (S) in slot until it just contacts the internal response lever on hydraulic pump, then tighten screw (S). Connect control rod and check "Response" Control operation.

214. INTERNAL LINKAGE ADJUSTMENTS. The following adjustments to internal lift linkage must be accomplished before installing lift cover. Special tools are required.

Invert the removed lift cover in a fixture or on bench, blocking cover so that rockshaft can be moved

to normal and raised transport positions. Move draft control link (D—Fig. 243) to fully raised position and make sure that draft control rod is against spring plunger by the linkage return spring. Clearance between adjusting screw (43) and lift cover should be 5.8 mm (0.230 inch) and can be measured using the correct size gauge as shown at (G).

215. To adjust draft control neutral setting for tractors **without intermix** (All except some 398 tractors), proceed as follows: Set the draft control link to "Sector" position and insert a 6.5 mm (1/4 inch) diameter pin (6—Fig. 244). Move position control link to "Transport", then insert a 6.5 mm (¼ inch) diameter pin (5). Loosen socket head screw (7) and lock nut on vertical lever to ensure that linkage does not interfere. Attach special tool (MF 273 or equivalent) to lift cover and attach a 1.3 kg (3 lbs.) load (9) to the end of vertical lever (10). The load (9) simulates the force of the linkage pump control spring. Attach special tool (MF 356C or equivalent) as shown at (11) to set the vertical lever to the correct (pump neutral) position. Loosen nut (29) and adjust position of lever pivot until lever just contacts the horizontal pin of special tool (11), then tighten nut (29). Recheck adjustment, then bend lock tab against nut (29). Refer to paragraph 217 for setting position control neutral ("Transport") setting.

216. To adjust draft control neutral setting for tractors **with intermix** (Some 398 tractors), proceed as follows: Set the draft control link to "Sector" position and insert a 6.5 mm (¼ inch) diameter pin (6—Fig. 245). Move position control link to "Transport", then insert a 6.5 mm (¼ inch) diameter pin (5). Loosen socket head screw (7) and lock nut on vertical lever to ensure that linkage doesn't interfere. Attach special tool (MF 445) as shown at (13) to correctly position the lift arms. Marks installed at the factory on the lift arm and lift cover should be aligned when the special tool is install. These marks on lift arm and

cover can be used to position the lift arms without using the special tool if carefully done and a retaining plate screw can be used to hold the arms in position.

Attach special tool (MF 273 or equivalent) to lift cover as shown at (8) and attach a 1.3 kg (3 lbs.) load (9) to the end of vertical lever (10). The load (9) simulates the force of the linkage pump control spring and a straight spring scale pull can be attached to lever (10) to apply the correct force.

Attach special tool (MF 356C or equivalent) as shown at (11) to set the vertical lever to the correct (pump neutral) position. Loosen nut (29) and adjust position of lever pivot until lever just contacts the horizontal pin of special tool (11), then tighten nut (29). Recheck adjustment, then bend lock tab against nut (29). Refer to paragraph 217 for setting position control neutral ("Transport") setting.

217. The following procedure for setting position control neutral setting is used for both tractors with and without intermix. Position the draft control link in fully UP position as shown at (6—Fig. 246). Draft control link does not have alignment holes in the fully UP position. Move position control link to "Transport," then insert a 6.5 mm (¼ inch) diameter pin (5).

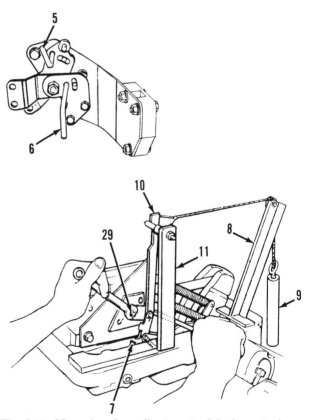

Fig. 244—View showing adjustment of draft control neutral setting for tractors without intermix (all except some 398 tractors). Special tool MF 273 (8) is used to attach load (9) to the end of vertical lever (10). Special tool MF 356C (11) sets the vertical lever to the correct (pump neutral) position.

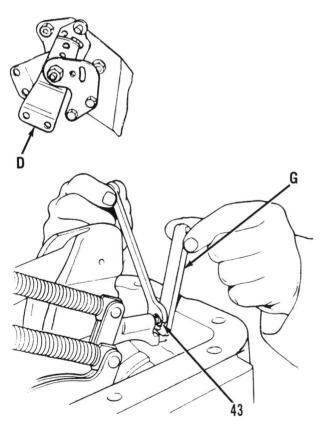

Fig. 243—Clearance between adjusting screw (43) and lift cover should be 5.8 mm (0.230 inch) and can be measured using the correct size gauge as shown at (G).

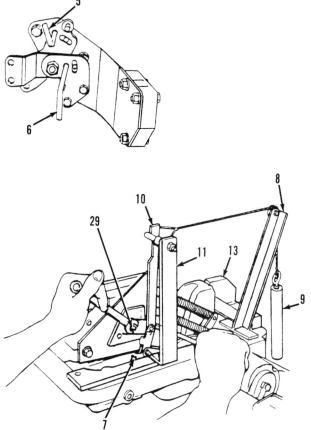

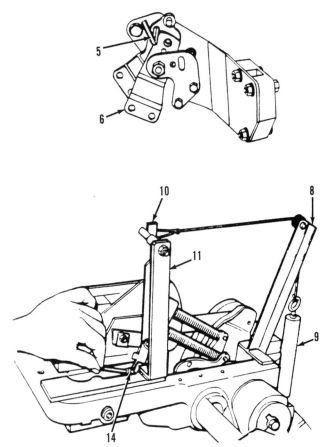

Fig. 245—View showing adjustment of draft control neutral setting for tractors with intermix (Some 398 tractors). Special tool MF 273 (8) is used to attach load (9) to the end of vertical lever (10). Special tool MF445 (13) can be used to set the position of the lift arms. Special tool MF 356C (11) sets the vertical lever to the correct (pump neutral) position.

Fig. 246—Draft control link (6) is fully UP in position shown. Position control link is in "Transport" location when pin (5) is installed through locating holes.

LIFT COVER

All Models

Rotate the cam ram arm against special tool (MF 272 or equivalent). If special tool is not available, marks (B and C—Fig. 241) can be aligned. Refer to paragraph 213 for correct placement of marks. Lift arm retaining plate screw can be used to hold the arms in position and simulate installation of special tool (MF 272).

Attach special tool (MF 273 or equivalent) to lift cover as shown at (8—Fig. 246) and attach a 1.3 kg (3 lbs.) load (9) to the end of vertical lever (10). The load (9) simulates the force of the linkage pump control spring and a straight spring scale pull can be attached to lever (10) to apply the correct force.

Attach special tool (MF 356C or equivalent) as shown at (11) then set position of the vertical lever (10) by turning screw at (14) until lever just contacts the horizontal pin of special tool (11), then tighten adjustment locknut. Recheck adjustment, after tightening locknut.

218. REMOVE AND REINSTALL. Disconnect lift links from lift arms and remove upper link control beam from master control spring clevis.

On models without cab, disconnect hydraulic lines attached to the spool valve block, disconnect spool valve control rods, then remove the spool valve. Disconnect the pto control rod to the left side cover, remove the four-wheel-drive control lever (if so equipped) and disconnect the response control rod from the right side cover. If so equipped, disconnect control lever and rod from right side cover. Disconnect draft and position control rods (7—Fig. 240) from control levers and detach wiring harness from under the seat. Unbolt seat panel from fenders and front vertical panel, loosen screws attaching fenders to rear axles, then lift complete seat panel and control levers from tractor.

On models with cab, disconnect spool valve control rods at front end, then remove the spool valve. Unbolt and remove the complete seat panel. Remove the

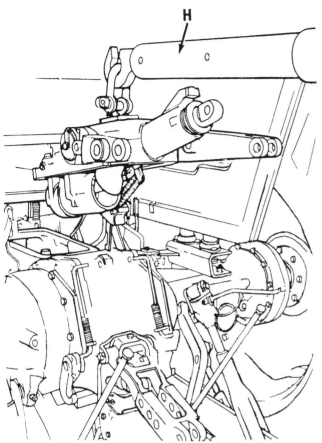

Fig. 247—Attach hoist (H) as shown for removing lift cover.

selector control valve lever and disconnect control rod from right side cover, if so equipped.

On all models, remove transfer cap and stand pipe or selector valve and stand pipe. Move draft control lever in full UP position as shown at (6—Fig. 246) and the position control lever in transport position and lock both levers in this position. Unbolt lift cover from housing and attach hoist to cap or sector valve as shown in Fig. 247.

Refer to proceeding paragraphs 214 through 217 for internal linkage adjustments. Clean sealing surfaces of center housing and lift cover and coat mating surfaces with "Loctite 515 Instant Gasket" or equivalent sealer. Install at least two 7/16 inch UNC studs approximately 2 inches (50 mm) long in center housing before installing cover to facilitate alignment. Be especially careful when lowering cover that vertical lever of cover does not catch on internal components and bend. Tighten retaining screws to 88 N.m (65 ft.-lbs.) torque. Refer to paragraphs 211 through 213 for external adjustments. Remainder of assembly is reverse of disassembly procedure.

219. OVERHAUL. It should be noted before disassembling the removed lift cover that special tools may be required to adjust linkage when reassembling.

Refer to paragraphs 213 through 217 for some adjustment notes.

To disassemble, loosen Allen set screw (9—Fig. 239) and unscrew adjusting nut (4), then withdraw parts (1 through 8) from housing (11). Release lock plate from around nut (29—Fig. 249), then remove nut, tab washer and spacer. Release spring (41), detach pivot bracket (31) from screw, then remove spring (38). Guide rods for control cam return springs (34) are drilled near loose end to assist disassembly and reassembly. Compress springs (34) and insert pins through rods to hold springs (34) compressed, then unbolt and remove ram cylinder (17) with bracket (30) attached. Always install new "O" ring (18) when assembling. Thread a 10-32 UNF screw into support retaining pin (7—Fig. 250), then pull or pry pin from lift cover. Hold control fingers (32 and 37—Fig. 249) away from cams (39 and 40) as support (3—Fig. 250) and shafts (1 and 2) are withdrawn. Cams (39 and 40—Fig. 249) can be removed after removing set screw and pivot pin. Rocker arms and shaft can be removed and new bushings installed if required.

Assemble lift cover using all new "O" rings and tab (locking) plates. Concave side of Belleville washer should be toward lift cover. When installing connecting rod (21) in rockshaft arm (26), coat threads of set screw (33) with "Loctite 542" or equivalent, tighten set screw until bottomed against connecting rod, then back screw up 1/4 turn. Tighten the two front nuts retaining cylinder (17) to 200-240 N.m (148-177 ft.-lbs.) torque and the two rear nuts to 280-330 N.m (207-243 ft.-lbs.) torque. Refer to paragraph 218 for installation and to paragraphs 213 through 217 for adjustment notes.

SELECTOR VALVE

Models So Equipped

220. Some models are equipped with a selector control valve (Fig. 252) which allows operation of lift linkage and auxiliary hydraulics simultaneously, or allows only the operation of auxiliary hydraulics. With the selector valve forward, in "LINKAGE" position, the lift linkage and auxiliary hydraulics may be operated independently. With selector lever back, in "EXTERNAL" (auxiliary) position, the flow of oil from the hydraulic lift pump and auxiliary hydraulic pump is combined to provide maximum flow at the remote couplers. The lift linkage is locked and hitch can not be used.

221. LINKAGE ADJUSTMENT. To adjust linkage, first move draft control and position control levers to "DOWN" position. Disconnect control linkage from selector valve bellcrank (7—Fig. 251). Rotate bellcrank clockwise until internal resistance is felt. Adjust length of lower control rod (10) by turning

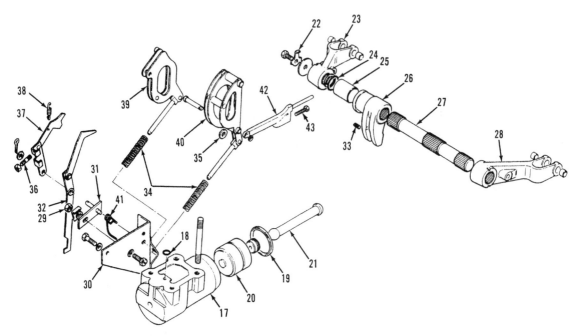

Fig. 249—Exploded view of typical hydraulic lift linkage, rockshaft and associated parts.

17. Cylinder
18. "O" ring
19. Piston rings
20. Piston
21. Connecting rod
22. Lock clip
23. Lift arm

24. "O" ring
25. Bushings
26. Rockshaft arm
27. Rockshaft
28. Lift arm
29. Nut
30. Support bracket

31. Pivot bracket
32. Vertical lever
33. Set screw
34. Springs
35. Cam roller
36. Position control
  adjusting screw

37. Position control finger
38. Spring
39. Position control cam
40. Draft control cam
41. Spring
42. Draft control rod
43. Stop screw

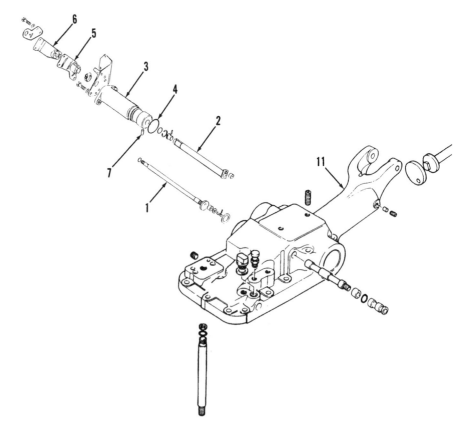

Fig. 250—Exploded view of typical control shafts, support and lift cover.

1. Draft control shaft
2. Position control shaft
3. Support
4. "O" ring
5. Position control lever
6. Draft control lever
7. Support retaining pin
11. Lift cover

clevis (8) until retaining pin (9) fits freely into clevis and control valve bellcrank, then lengthen the rod by one turn of the clevis and install the pin.

Turn adjusting screw (6) as required so that lever just snaps into position and requires slight force. Tighten locknut when adjustment is correct.

222. R&R AND OVERHAUL. To remove the selector valve, disconnect control linkage and hydraulic pipes from valve, then remove both socket head screws retaining valve to lift cover. Lift valve from lift cover.

To disassemble, unbolt and remove end plate (30—Fig. 252). Remove dowel (D), then remove actuator shaft (26) and fork (29). Spool (28) can be removed from cover end, but plug and spring should also be removed. Remove plugs, shims (22), springs and relief valve (23). Relief valve is used to protect lift cylinder from shock loads when transporting heavy mounted implements over rough terrain with valve in external (auxiliary only) position.

Clean all parts and be sure that all old "O" rings and gaskets are removed. Clean all oil passages using compressed air.

Reassemble selector valve, reversing disassembly procedures. Relief valve (23) should open at 262-278 bar (3800-4000 psi). Increasing spring pressure by adding 0.25 mm (0.010 inch) thickness to shims (22) will increase pressure approximately 7 bar (100 psi). Relief valve opening pressure can be checked using appropriate test fixtures and pressure source above the test range.

Reinstall selector valve using new gasket and "O" ring. Tighten retaining socket head screws to 34-54 N•m (20-40 ft.-lbs.) torque. Adjust control linkage as outlined in paragraph 202.

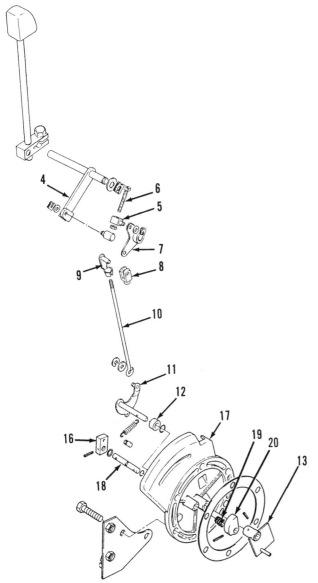

Fig. 251—View of right side cover (17) and linkage. Bellcrank (7) is also shown in Fig. 252.

4. Control lever	
5. Pivot	12. Bushing
6. Adjusting screw	13. Arm
7. Bellcrank	16. Response control lever
8. Clevis	17. Side cover
9. Retaining pin	18. Shaft
10. Lower control rod	19. Spring
11. Lockout lever	20. Response cam

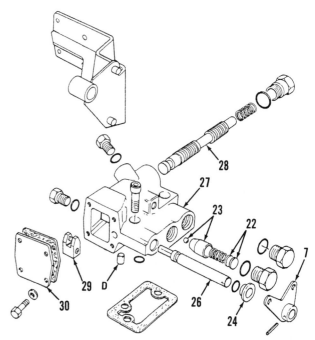

Fig. 252—Exploded view of selector valve assembly used on some models. Bellcrank (7) is also shown in Fig. 251.

7. Bellcrank	26. Actuator shaft
22. Shims	27. Selector valve body
23. Safety relief valve	28. Valve spool
poppet and ball	29. Shift fork
24. Sleeve	30. End plate

HYDRAULIC LIFT PUMP

All Models

223. REMOVE AND REINSTALL. The lift pump can be removed from models **equipped with only single speed (540 rpm), ground speed pto**, through the top opening of differential housing after removing the lift cover. However, many mechanics prefer to split the tractor between transmission and differential housing, then remove the pump out front opening of differential housing as outlined in paragraph 224.

To remove the lift pump from tractors which can have pump removed through the top, drain transmission oil and remove lift cover as outlined in paragraph 218. Remove the pto output shaft and side covers from both the left and right sides of axle center housing. Remove split pin from center of drive shaft coupling, then remove the coupling and drive shaft. Unbolt and remove the two retaining screws or stud nuts and the pump supports (S—Fig. 253) from each side of axle center housing, then carefully work pump assembly up out of center housing.

Install by reversing removal procedure. Be sure that all old gaskets and "O" rings are removed completely and surfaces are cleaned. Use new gaskets and "O" rings when assembling. Install new split pin in drive shaft coupling hole that limits drive shaft end play to 0.38-2.54 mm (0.015-0.100 inch).

224. To remove the lift pump from most tractors (except those equipped with only single speed (540 rpm), ground speed pto, it is necessary to drain transmission oil and remove lift cover as outlined in paragraph 218. Split tractor between differential housing and transmission housing as outlined in the appro-

priate paragraph in the transmission section. Remove pto output shaft and cover from right side of rear axle center housing. Disconnect hydraulic lines and remove pump supports (S—Fig. 253), then remove hydraulic pump unit through front of differential housing.

Installation is reverse of removal procedure. Always install new split pins, "O" rings and gaskets. Carefully locate the pump on ipto clutch and install supports (S). Tighten nuts or screws retaining support (S) to right side, but leave left side support retaining nuts or screws loose until tractor assembly is complete. Be sure that engine will not start and crank engine with starter, then tighten nuts or screws retaining support to left side of axle center housing. This procedure will ensure correct alignment of shafts.

225. OVERHAUL. To disassemble the removed pump, first clean the outside of pump, then unbolt and remove the pressure relief valve assembly (1—Fig. 254). Remove clip (2) and lever (3). Remove the four lower screws (4) retaining cover (7), then separate inlet housing (17) from front section of pump. Remove the remaining two screws (6) and separate rear cover (7) from inlet housing. Remove clip (9), nut (10), spring (11), washer (12), "O" ring (13), filter (14), inner shroud (15) and upper "O" ring (16). Unhook spring (18), remove screws (19), then remove response control lever (20) and retainer (21). Ball, spring and bushing (24) can be removed after removing response lever. Compress spring and remove retainer snap ring (25), retainers (26 and 28), spring (27), seal (29) and valve (30).

Remove cap (31—Fig. 255), compress spring (37) using MF352 retainer clip or equivalent and remove retaining ring (32). Withdraw control valve (33) and balancer tube (34). Be careful that spring (37) is not ejected and remove special tool MF352. Remove collar (35), guide (36), spring (37) and square disc (38). Remove four retaining nuts (39), then withdraw front cover (40) from studs. Remove front thrust washer (42) and "O" rings (41). Slide valve chambers (43) from studs withdrawing pistons (44 and 50), cam blocks (45 and 49) and oscillator drive (46). Carefully separate parts and remove camshaft (51) from rear body (52). Control valve body (53) can be driven from rear body if new "O" rings and back up rings are to be installed. Special tool (MF351 or equivalent) can be used to remove valve assemblies from valve chambers (43) if service is required.

When assembling, back up rings are outside "O" rings at positions (54). Special tool (MF353 or equivalent) can be used to install back up rings and "O" rings (54) on control valve. Align pin for control valve (53) when installing and use special tool (MF354 or equivalent) to drive valve into rear body bore.

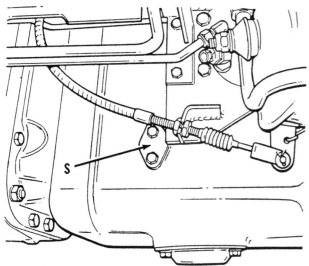

Fig. 253—Hydraulic pump supports (S) are attached to each side of rear axle center housing.

Fig. 254—Rear view of partially disassembled hydraulic lift pump filter, rear cover and inlet housing. Refer to Fig. 255 for pump parts at front.

1. Pressure relief valve & housing
2. Clip
3. Actuating lever
4. Screws (4 used)
5. Gasket
6. Screws (2 used)
7. Rear cover
8. Gasket
9. Clip
10. Nut
11. Spring
12. Washer
13. "O" ring
14. Filter
15. Inner shroud
16. "O" ring
17. Inlet housing
18. Spring
19. Screws (2 used)
20. Response control lever
21. Retainer
22. Bushing
23. Washer
24. Bushing, spring & ball
25. Snap ring
26. Retainer
27. Spring
28. Retainer
29. Seal
30. Valve

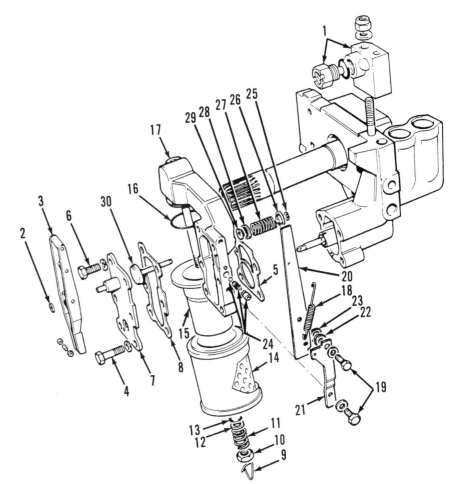

Valve seats can be recut in chambers (43) using special tool (MF349 or equivalent). Special tools (MF350, MF351 and MF353) can be used to assemble valves and valve chambers.

Reassemble by reversing disassembly procedure. Lubricate all parts liberally before assembling and always install new "O" rings and retaining rings. Chamfered corners (C) of piston yokes face away from each other. Tighten nuts (39) to 40 N·m (30 ft.-lbs.) torque. If only two of the nuts (39) are special type, install these two on right top and bottom left studs. Assemble balancer tube (34) to control valve (33), lubricate valve and insert into rear body.

Tighten nut retaining relief valve manifold (1—Fig. 254) to 27 N·m (20 ft.-lbs.) torque. Tighten screws (4 and 6) evenly to 20 N·m (15 ft.-lbs.) torque, while checking to be sure that control valve (33—Fig. 255) continues to move freely. After clip is fitted to pin (47), bend end over to prevent its removal.

AUXILIARY HYDRAULIC PUMP

All Models So Equipped

226. The power steering pump of most models is tandem or dual type, incorporating one section for power steering and the other section for the auxiliary hydraulic system. Refer to paragraph 25 for removal and to paragraph 30 for overhaul.

AUXILIARY SYSTEM PRESSURE

All Models So Equipped

227. A pressure maintaining valve (5—Fig. 256) is fitted to tractors with ipto or Multi-Power transmission and is used to maintain sufficient pressure to operate these clutches. Valves with different maintaining pressures are used for tractors with three cylinder engines (17-20 bar, 250-290 psi) than for tractors with four or six cylinder engines (19-22 bar, 275-325 psi). Pressure is stamped on valve body and should be correct for installation. Pressure is not adjustable, but can be checked at port (8) with oil warm and engine operating at 1200 rpm. System pressure may be increased by clogged hydraulic system filter, so new filter should be installed before checking pressure. Low pressure could be caused by:

1. Faulty pressure maintaining valve.
2. Steering pump worn
3. Leakage of ipto system caused by loose connection or ipto brake cylinder leaking.

4. Leaking Multi-power clutch pack.

228. Auxiliary system pressure relief valve (Fig. 257) should maintain auxiliary system pressure at 170-190 bar (2500-2750 psi) for all models. Pressure

can be checked by attaching appropriate test gauge in place of port plug (1) located in manifold mounted just in front of left rear axle housing. Start engine and make sure fluid is warmed. Be sure that selector valve is in linkage position and set engine speed to

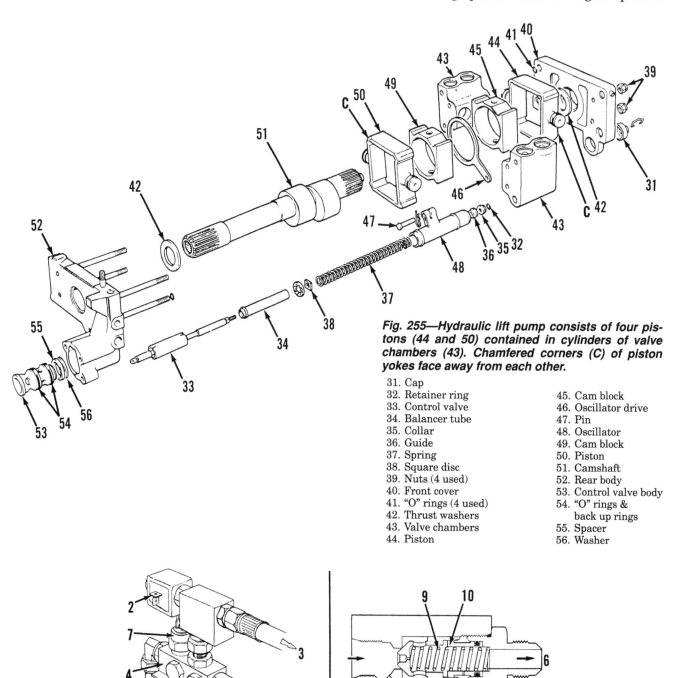

Fig. 255—Hydraulic lift pump consists of four pistons (44 and 50) contained in cylinders of valve chambers (43). Chamfered corners (C) of piston yokes face away from each other.

31. Cap	
32. Retainer ring	45. Cam block
33. Control valve	46. Oscillator drive
34. Balancer tube	47. Pin
35. Collar	48. Oscillator
36. Guide	49. Cam block
37. Spring	50. Piston
38. Square disc	51. Camshaft
39. Nuts (4 used)	52. Rear body
40. Front cover	53. Control valve body
41. "O" rings (4 used)	54. "O" rings &
42. Thrust washers	back up rings
43. Valve chambers	55. Spacer
44. Piston	56. Washer

Fig. 256—Views of installed Multi-Power solenoid (2) and pressure maintaining valve (5). Cross section shows internal components of pressure maintaining valve. Models without Multi-Power will not be equipped with solenoid (2), block or outlet at (3).

1. Pressure from steering system	4. Manifold	6. Pressure to auxiliary hydraulic pump	9. Spring
2. Multi-Power solenoid	5. Pressure maintaining valve	7. Low pressure switch	10. Dampening shuttle
3. Pressure to Multi-Power		8. Test port	11. Poppet
			12. Orifice

1200 rpm. Pull one of the auxiliary control valve spools back to lift position and note pressure indicated by the installed test gauge. Low pressure could be caused by:

1. Auxiliary pump worn.
2. Relief valve faulty.
3. Internal leakage across spool valve.

Low relief valve setting can be corrected by removing plug and adding shims (3) as required. Be sure to install new seals and tighten relief valve plug to 70-100 N•m (52074 ft.-lbs.) torque.

229. Auxiliary system flow can be checked by attaching an appropriate flow tester to outlet port connection. Direct fluid from tester back to the transmission reservoir through filler opening. Start engine, make sure that flow meter restrictor valve is open and allow fluid to reach normal temperature. Make sure that ipto is in "OFF" position and set engine speed to 2000 rpm. Slowly close flow meter restrictor valve until pressure increases to 138 bar (2000 psi) and observe volume of flow. Open restriction as soon as flow is determined. New pumps on four cylinder tractors should deliver 32 L/minute (8.4 gal./minute) or at least 28 L/minute (7.4 gal./minute) for used pumps. Low volume of flow could be caused by worn pump, blocked (or partially blocked) suction strainer or insufficient amount of oil in transmission sump.

AUXILIARY REMOTE CONTROL VALVES

All Models So Equipped

230. One, two or three auxiliary spool valves may be installed on bracket at rear of tractor. Cap parts (A—Fig. 258) are for valves with detent positions which return (kickout) to neutral when cylinder reaches end of stroke. Cap parts (B) are for valves with spring return to neutral. Cap parts (C) provide

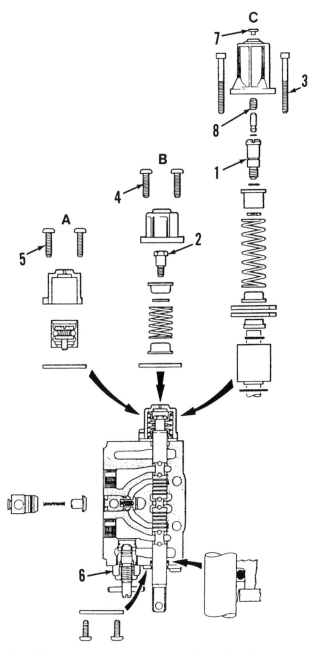

Fig. 258—Exploded and cross section of auxiliary spool control valve. Cap parts (A) are for valves with three position detent, cap parts (B) are for valves with spring return to neutral and cap parts (C) are for valves with detent type with float position.

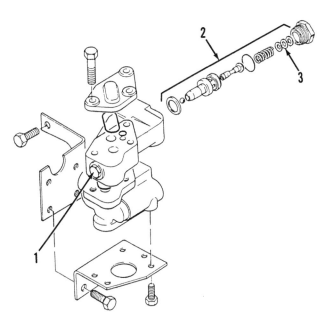

Fig. 257—Exploded view of relief valve (2) for auxiliary hydraulic system. Pressure can be checked at port for plug (1). Shims (3) are located between spring and plug.

four positions, including detent and float positions. Additional special valves are also available including type without kickout and valve for use with hydraulic motors. Service is conventional and should be accomplished if seals and repair parts are available. "Loctite 542" or equivalent should be applied to threads of shoulder bolt (1 or 2) and tighten shoulder bolt to 7 N·m (5 ft.-lbs.) torque. Apply "Loctite 222" to threads of screws (3, 4 or 5) and tighten to 7 N·m (5 ft.-lbs.) torque. Tighten change-over valve (6) to 22 N·m (17 ft.-lbs.) torque.

To adjust kickout relief pressure of valves (C), attach inlet hose for flow meter to valve and connect outlet side of flow meter to the other port of valve. Remove plug (7), open flowmeter fully and start engine. After fluid reaches normal operating temperature, set engine speed at 2000 rpm and move valve control to locked position. Close the flow meter restriction valve slowly until pressure causes valve to kickout (return to neutral). Kickout pressure should be 172 bar (2500 psi) and can be changed by turning adjusting screw (8) with a 5/32 inch Allen wrench through opening for plug (7). Turning screw clockwise will increase kickout pressure and each 10° will change pressure about 2 bar (30 psi). Check setting several times before reinstalling plug (7).

NOTES

NOTES